# Introduction to Systems Analysis

# Introduction to Systems Analysis

**GERALD A. SILVER**
*Los Angeles City College*

**JOAN B. SILVER**

Prentice-Hall, Inc., Englewood Cliffs, New Jersey

*Library of Congress Cataloging in Publication Data*

SILVER, GERALD A.
Introduction to systems analysis.

1. Electronic data processing—Business.
2. System analysis. I. Silver, Joan B., joint
author. II. Title.
HF5548.2.S4494 658.4′032 75–15554
ISBN 0–13–498683–0



Printed in the United States of America

10 9 8 7 6 5 4 3 2 1

PRENTICE-HALL INTERNATIONAL, INC., *London*
PRENTICE-HALL OF AUSTRALIA, PTY. LTD., *Sydney*
PRENTICE-HALL OF CANADA, LTD., *Toronto*
PRENTICE-HALL OF INDIA PRIVATE LIMITED, *New Delhi*
PRENTICE-HALL OF JAPAN, INC., *Tokyo*
PRENTICE-HALL OF SOUTHEAST ASIA (PTE.) LTD., *Singapore*

# Contents

# Preface

The pace and pressures of modern business activities have ended forever the places of chance, tradition, and guesswork in the decision-making process. The very life of a modern business enterprise depends on the availability of accurate, relevant data processed economically and promptly.

The need for systems analysis to develop and guide this life-giving data flow has increased phenomenally and continues to expand into many other business areas as well. The future of business systems analysis shows great promise. The economic rewards from the systems approach to solving business problems are clearly evident. Large and small businesses alike are feeling the necessity of developing business systems in order to compete more successfully in today's marketplace.

Many tasks face today's systems analysts. They must be capable of solving a wide range of complex and sophisticated business problems. They must be able to manipulate and guide the expenditure of large sums of money, supervise the selection and installation of expensive, sophisticated electronic equipment; alter the physical environment of the business plant; train or retrain hundreds of employees; and cope with the psychological reactions of customers, employees, and management.

The costs involved in modern business systems have surged up in the last several decades. Systems must be implemented as nearly perfectly as possible the first time around—there is little room for second chances, trial and error, or experimentation.

This text introduces the undergraduate student to the world of systems analysis. It defines the role of systems analysts and describes their duties, methods, techniques, and tools. It provides a comprehensive and complete survey of the major aspects, concepts, and theories of business systems. It stresses the fundamental while reducing extraneous detail. It strikes a balance between theoretical and applied phases of the discipline.

The text incorporates a variety of approaches designed to maximize its instructional value to the student. It presents fundamental concepts clearly and concisely in nonmathematical terms. It includes an explanation and discussion of the principles of the scientific method and encourages the student to rely on these principles as a sound basis for all problem-solving activities.

The text draws liberally upon visual media—pictures, drawings, flowcharts, and diagrams—to illustrate concepts and to reinforce the development of a sense of basic logic and reasoning.

In addition, the text stresses modern theory and applications, including electronic data processing and computers. Teleprocessing and communications theory are discussed and integrated into many of the examples.

Another important aspect covered in this text is the thorough treatment of word processing systems. Born as the "little brother" of data processing, word processing has come of age. The latest concepts related to text manipulation, as well as word processing, are explored with both theory and examples. Modern methods of data input, output, storage, and retrieval are also presented, along with a thorough discussion of all microforms—microfilm, microfiche, aperture cards, computer output microfilm, and others.

Finally, the text presents a comprehensive group of case histories. Drawn from real life situations, these diverse cases explore the wide variety of business systems found in industry today. They expose the student to the type of situations and problems that will be encountered on the job. The case histories are presented in survey form to allow the student to integrate the knowledge gained from the text without being burdened by less important details.

The authors would like to thank the manufacturers and equipment vendors who supplied the photographs and illustrations presented throughout the text.

*Gerald A. Silver*
*Joan B. Silver*

# Chapter 1

# The Role of Business Systems Analysis

To some, this is the atomic age; to others, the computer age. But perhaps it might better be called the paperwork age.

Paperwork is the most abundant product produced by modern business. And processing this paperwork occupies the working hours of hundreds of thousands of secretaries, clerks, managers, office workers, and others.

Why the volume of paperwork? Basically, for two reasons—people and government.

Each person throughout his or her lifetime becomes the subject of an enormous amount of information—medical files, insurance policies, financial records, mortgage loans, employment and school records, to name a few. A growing population generates a tremendous amount of data and paperwork.

The second cause of the multitude of paperwork is increasing governmental regulations and control. More and more, records of transactions that make up daily business life must be preserved in the form of reports and documents for governmental perusal and processing.

In the year 1967, for example, it cost private industry $20 billion just to generate the data required by the government alone. The federal government spent another $8 billion to process its way through the mountain of paperwork created.[1] And this represents only a part of the paperwork produced

[1] Rudolph J. Passero, "Better Utilization of Systems," *The Office* (January 1972), p. 104.

throughout the nation by large and small businesses, schools, hospitals, factories, and others.

Data is used by different organizations in many different ways and for different purposes. It becomes the information base from which large and small business decisions are made. It is used to set policy and to dictate employment, purchasing, and distribution patterns. More often than not, it creates even more data.

The flow and availability of data has become an indispensable, vital element of almost every human enterprise. Molding this data into coherent, usable form does not occur by chance. It requires planning, people, machines, and money.

Over the last several decades the science of systems analysis has developed to meet this need. Its goal is to make the appropriate data available

**Figure 1.1.** Charts

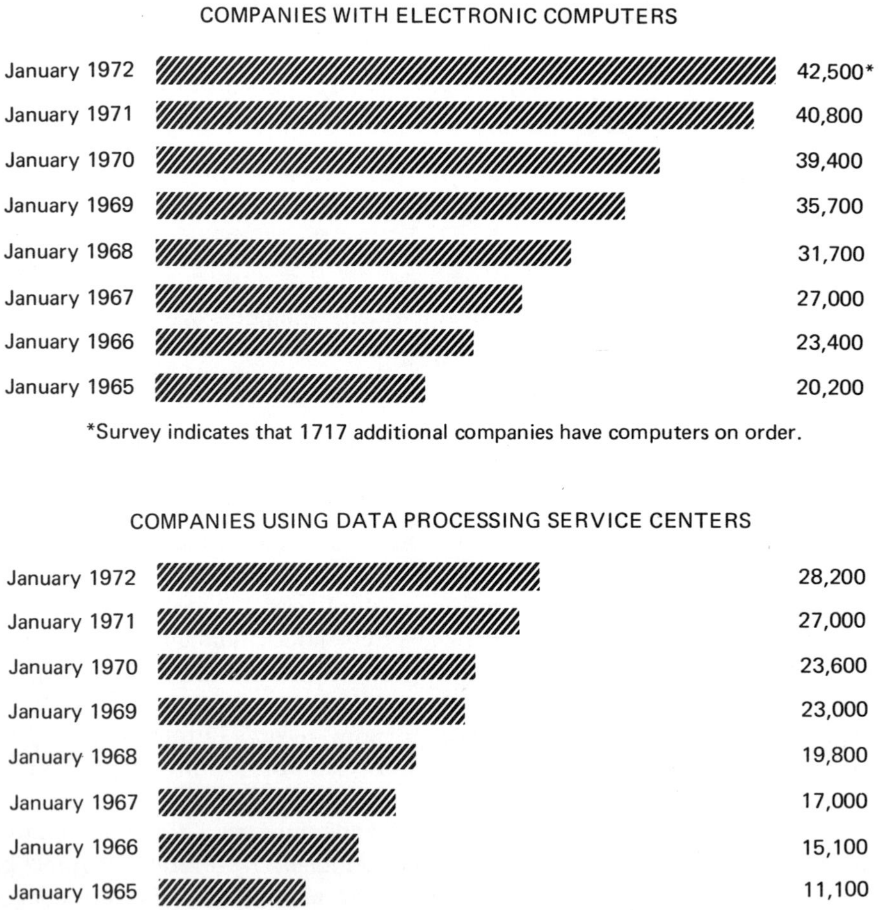

COMPANIES WITH ELECTRONIC COMPUTERS

| | |
|---|---|
| January 1972 | 42,500* |
| January 1971 | 40,800 |
| January 1970 | 39,400 |
| January 1969 | 35,700 |
| January 1968 | 31,700 |
| January 1967 | 27,000 |
| January 1966 | 23,400 |
| January 1965 | 20,200 |

*Survey indicates that 1717 additional companies have computers on order.

COMPANIES USING DATA PROCESSING SERVICE CENTERS

| | |
|---|---|
| January 1972 | 28,200 |
| January 1971 | 27,000 |
| January 1970 | 23,600 |
| January 1969 | 23,000 |
| January 1968 | 19,800 |
| January 1967 | 17,000 |
| January 1966 | 15,100 |
| January 1965 | 11,100 |

Source: *The Office*, "Progress in Automation Made by Readers of *The Office*," Jan. 1972, p. 22, Vol. 75, No. 1.

to the right person at the right time in the business cycle, and at a cost and level of accuracy suitable to the needs of the organization. The job of the systems analyst has grown in stature and number, becoming an indispensable part of modern business operations.

## BUSINESS SYSTEMS—PAST TO PRESENT

The demand for data has not always been so great. It reflects the tempo, temperament, and technology of the times. Before the advent of jet airplanes, intercontinental television, and international telephone networks, the pace of life and business was slower and more leisurely.

Travel was measured in days, not hours. One could not hop a plane in the morning, arrive in another city by noon for an important meeting, and be home again in time for dinner.

The telephone was fine for conducting local business, but most communications beyond that distance went by mail or telegram. And since the shipment of goods and supplies was expensive and slow, business dealings with companies outside the local area were relatively infrequent.

Radio, newspapers, and magazines were slowly breaking down the pro-

**Figure 1.2.** Business office environment circa 1920

Courtesy of the Bureau of the Census library.

vincialism and isolation of the country, but there was no national television to infuse new ideas, new products, and new methods into everyone's living room in full color.

The attitudes and atmosphere of business firms 40 to 60 years ago reflected these conditions. Businesses were small, local, individualized, and often family-owned. The owner, or proprietor, served in many capacities, without any academic training for many of the tasks he or she performed.

Information gathering and data processing activities were limited to what the proprietor might need in the short term—for example, hours worked by employees or amounts owed by credit customers. Most data processing had to be done by hand, and it was relatively time-consuming and expensive.

The quality of business decisions reflected this lack of precision. Most business people relied on hunches, guesses, intuition, personal experience, or plain old tradition to solve business problems. Since the influence of a single decision, made by one person, was limited—usually involving only a few people and a small amount of money—these methods often appeared to be sufficient. Consequently, errors or mistakes in judgements were also limited in their effects, and, most of the time, changes and corrections were easily made.

Attitudes toward business problems reflected the times as well. The time pressures that we know today, and the far-reaching effects of many business decisions, were not yet present.

**Figure 1.3.** Cataloging system, early 1900's

Courtesy of the Bureau of the Census library.

**Figure 1.4.** Early keypunch equipment from the 1900's

Courtesy of the Bureau of the Census library.

Forty or sixty years ago, if a businessman wanted to improve the data processing activities in his office, he might wait until a lucky salesman dropped in to show off the latest in manual typewriters or mechanical calculators. If he had the money, and it seemed practical, he made a purchase on the spur of the moment. On the other hand, if a salesman did not come around and the problem could not be solved immediately—well, it would keep for a while until more time could be devoted to it.

This method of solving problems was, in some ways, adequate for the times. For most small problems the range of available solutions was limited and usually within the knowledge and grasp of the ordinary business person or manager. And, as yet, there were not a great many alternative methods for solving business problems.

But times were changing—the population was growing, technology was advancing. Business was on the verge of outgrowing many of its traditional, leisurely ways.

World War II thrust the nation into a new era. Industry was suddenly faced with the monumental task of supplying the country with the tools of defense and war. These new tools were more complicated than any that had yet been manufactured, and they were needed faster than ever before and in ever larger quantities.

With the nation's survival at stake, industry could not rely on guesswork or inefficient methods of decision making. Methods lacked precision and reproducibility. They had high error factors, no cost controls, and relied

**Figure 1.5.** Early data entry device used by the Bureau of the Census

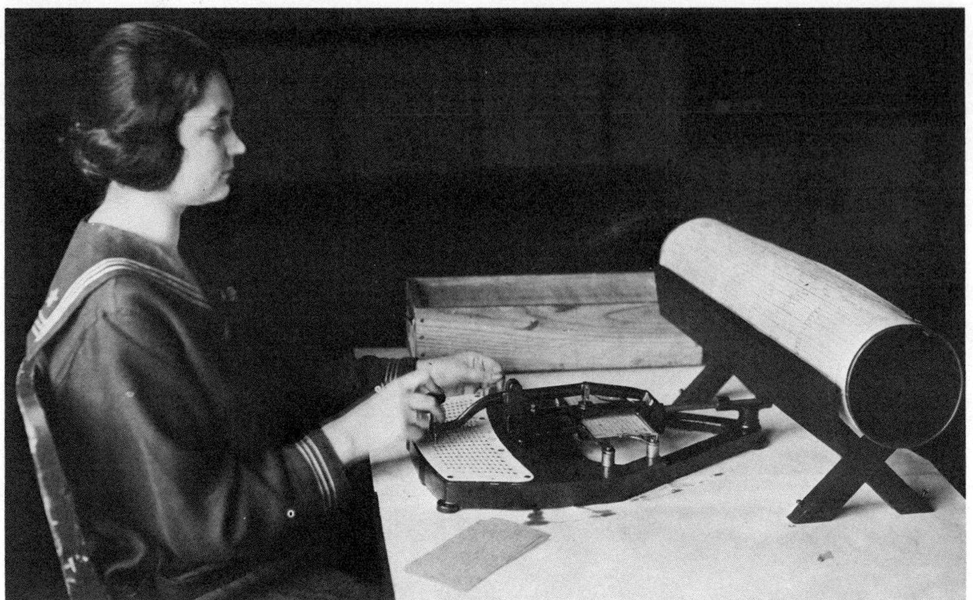

Courtesy of the Bureau of the Census library.

too much on isolated individuals rather than on systems. Managers, now under pressure, looked for ways to improve efficiency and output. Scheduling, timing, and planning were critical if defense plants were to meet the demands placed upon them.

The scientific method of problem solving and statistical techniques pointed the way toward the answers. These tools gave managers powerful systematic procedures for analyzing, defining, and evaluating new techniques and solutions to problems. They were used to develop programs that could find the best mixture of employees, machines, and materials to produce the maximum output in the least amount of time. Terms such as PERT (Program Evaluation and Review Technique), operations research, systems analysis, and CPM (Critical Path Method) were born of this thrust.

With the birth of the computer in the late 1940's, the fledgling systems analysts were quick to recognize an even more powerful tool. The computer joined their growing repertoire.

## ACCOUNTABILITY AND COST EFFECTIVENESS

As time passed, other factors increased the pressure on the decision-making process. In the 1960's and 1970's, inflation, pressure on government to reduce expenditures and taxes, and international competition forced business people and government administrators to look carefully at costs. Now managers not only had to find the most efficient ways to produce products; they also had to account for each dollar spent.

The emphasis shifted to accountability and evaluating results in terms of costs. Such cost effective programs involve justifying each dollar spent with respect to outcomes or benefits returned.

Business and industry turned to their systems analysis departments to structure their cost effectiveness programs. Systems analysts recognized that such a company program was largely dependent on data gathered from all parts of an organization, processed, and then dispersed in new form back again throughout the firm. Designing an effective, efficient data flow system became the primary step in structuring a cost effectiveness program.

To do the job, systems analysts at first borrowed cost reduction and planning tools and techniques from production and manufacturing departments. As their stature and experience grew, they began to design and develop their own tools. They drew upon many disciplines—statistics, electronic data processing, psychology, management, and more. Today the systems analyst uses many sophisticated, specialized, and effective tools to design better and more efficient ways of moving data throughout an organization.

## FUNCTIONS OF THE MODERN BUSINESS SYSTEMS DEPARTMENT

Thousands of business firms across the country now employ trained staffs of systems analysts, and the number will continue to grow. In the early 1960's, about 10,000 people were formally employed as systems analysts. By the end of the decade there were over 200,000 individuals in this category. By the end of the 1970's, the demand for systems analysts will increase by 183 percent.[2]

Systems analysts in a firm often work in a centralized unit called the systems analysis or operations research department. Sometimes their function falls elsewhere on the organization chart. They may be part of the data processing department, under the direction of the company controller, or within the scope of the accounting department.

While the precise role of the systems analyst is not as yet formally defined, most serve in similar capacities within a firm. However, specific duties may vary depending on where this position falls in the organizational structure, the size of firm, and other factors. Analysts are usually involved in the following areas:

1. Financial and accounting policy decisions. Systems analysts assist management in setting corporate policies regarding financial and accounting matters and objectives.

2. Forecasting and simulation. Systems analysts play an important role in analyzing and planning future goals and growth patterns of a company.

[2] William J. Rost, "A Quarter-Century of Systems," *The Office* (January 1972), p. 60.

**Figure 1.6.** Business systems circa 1975

Courtesy Control Data Corp.

They predict financial and economic trends using various mathematical and computer techniques.

3. Sales and marketing of goods and services. Systems analysts supply information and research data and assist in decisions regarding the sales, marketing, and distribution of a firm's goods.

4. Planning the orderly flow of data throughout an entire business enterprise. Analysts specify what data will be collected, what form it will take, and how it will be processed and reported to management.

5. Modifying or redesigning existing business systems. The systems analyst has the responsibility for proposing changes and improvements in a data flow system to meet new or continuing needs of a business firm.

6. System implementation. Once a system has been designed and specified, the systems analyst directs and oversees its orderly implementation. This involves such factors as the selection of personnel and equipment, and the hiring of consultants.

7. Computer programming and utilization. Systems analysts indicate the types of computer programs needed for a new system, and monitor their development, testing, and final implementation. (The actual task of writing programs is often delegated to a programmer.)

8. File design and maintenance. Systems analysts specify what information will be included in a file and plan the layout of data fields in the

records. They define and develop the procedures necessary to keep the information in the file current and accurate. This includes processes such as adding new records or deleting outdated records.

9. Forms design and management. Systems analysts plan and design the forms, documents, and records to be used in a system. They select the forms, their content, sizes, routing, and storage, and specify maintenance procedures.

10. Establishing policies and procedures. Systems analysts write the policy manuals that are necessary for a system to work efficiently, and they define routines and methods for handling data flow.

11. Employment and training of company personnel. Systems people are involved in specifying personnel needs for various clerical and office work stations, and they describe training programs and procedures.

12. Work measurement. The system analyst records data on the output of various clerical, management, accounting, and data processing personnel. This is used to compare, report, and analyze the productivity of personnel involved in the flow of paperwork.

13. Work simplification. Analysts develop easier, faster, more accurate, and improved methods of processing data.

14. Office layout. Systems analysts are involved in planning the office environment and many working conditions throughout the firm. They help to design and lay out offices and administrative suites, including such things as the number and type of desks, cabinets, files, or dictation equipment. The analyst indicates the number of work stations, their locations, and the equipment needed by each.

15. Selection and specification of office and data processing equipment and supplies. A substantial amount of time is spent by the analyst in comparing and selecting hardware, devices, and other machines necessary for the efficient movement of data throughout a firm. This includes computers, unit record machines, copying machines, duplicating equipment, mail handling, routing machines, and others.

16. Planning and designing internal and external communications. An important responsibility of the systems analyst involves the firm's data and electronic communications facilities. This includes specifying the type and number of transmission lines, data lines, switch-boards, dictation equipment, telephones, Teletype terminals, data transmission equipment, and facsimile machines.

## ADVANTAGES OF USING SYSTEMS ANALYSIS

During the past three to four decades, the application of systems analysis techniques to business problems has benefitted businesses in several ways. This method of problem analysis and solution has shown results in these key areas of the business operation:

1. Efficient systems. Systems analysis methods help a firm to develop and maintain an organizational structure and operating procedures that lead to attaining maximum efficiency.

2. Maximizing profits. A business firm that operates efficiently and systematically is more likely to generate greater profits.

3. Resources used to the best advantage. Systems analysis aids a firm in achieving maximum output with the least investment of time, material, manpower, and other resources.

4. Reduction of human effort. Systems analysis encourages the best utilization and allocation of human effort and labor. Systems analysts attempt to uncover duplication, redundancy, and wasted effort, and to automate procedures whenever practical.

5. Speedier delivery of the end product. Effectively organized procedures and operations make it possible to reach end goals faster. This is true whether the output of the firm is the production of a good or service, or the movement of information, data, money, or the like.

6. Reducing or eliminating errors in data and information. An important goal of the systems analyst is to increase the level of accuracy of information generated and processed by a firm.

7. Documenting methods and procedures. Clearly written policy statements, diagrams, flowcharts, and such make it more likely that a firm's procedures and practices will be followed and maintained in a consistent manner. They also serve as guides for needed modifications or alterations.

## LIMITATIONS OF SYSTEMS ANALYSIS

Systems analysis is not a panacea for all business problems. It has limitations and weaknesses that must be understood and considered during all phases of systems analysis work. The following list summarizes some of these areas.

1. Some business problems are beyond the scope of systems analysis techniques. The most skillfully designed data flow system cannot help a company that is failing because of serious financial problems or because it is marketing a product no longer wanted by society. Problems created by pressures from outside the enterprise, such as from stockholders or the public, are also often beyond the grasp of the systems analyst.

2. Systems analysis costs money, time, and effort. Finding a long-term, permanent solution to a problem can be an expensive investment, compared to a short-term temporary answer.

3. Human elements can cause complications. A systems analyst must take the human element into consideration in his or her work. Much of the analyst's activities involves making changes—in routines, systems, organizational structure, working patterns and conditions, and other areas.

People generally oppose changes. They resist adjusting to unfamiliar situations, even to those that are in their best interests in the long run. The systems analyst has the responsibility of preparing employees, customers, management, vendors, and others to accept modifications and alterations introduced by a new system.

4. Acceptance takes time. No matter how promising a systems analysis project may seem, it cannot "sell" itself. Very often it must be "sold" to the people involved. They must be encouraged to cooperate in the development and implementation of the venture to insure its success.

## CHARACTERISTICS OF A SYSTEM

"System" is an overworked word in the English language. It is used to describe many things, many conditions, and many methods. We speak of an accounting system, a communications system, a circulation system, a transportation system.

What does the word "system" mean, and what are its attributes? Webster defines "system" as a regularly interacting or interdependent group of items forming a unified whole. Other definitions are: an integrated whole; a collection of related elements treated as a unit.

Each day we come into contact with many different types of systems. A city's water system, for example, is made up of pipes, valves, pumping equipment, and reservoirs, as well as its operating personnel and management. The climate in a given region is a result of the meteorological system. The plants and animals are elements of the ecological system. The human body is a system composed of many smaller systems. The government operates a tax collection system, and the courts are part of the legal system. Business, too, has its systems—accounting, payroll, quality control, and others.

One major attribute of the system concept has important implications for systems analysis—a system often possesses qualities or capabilities not found in the elements that compose it. This attribute—the whole producing results that are greater than the sum of the parts—is the key to understanding the business system concept.

To illustrate: Imagine a miscellaneous collection of mechanical and electrical items—bearings, gears, wires, nuts, bolts, stamped-out parts. Each part has a limited function and is of relatively little value by itself. But if all these parts are assembled into a luxury automobile, the miscellaneous collection becomes an integrated, functioning system with practical and monetary value. The relative worth of the resulting system is far greater than the sum of its parts.

The same is true of a business system. Office machines, personnel, procedures, invoices, and reports functioning as an integrated system may have a far greater value than have all the parts by themselves.

A business system, like any other system, is a unified whole composed of

Figure 1.7. A business system and its elements

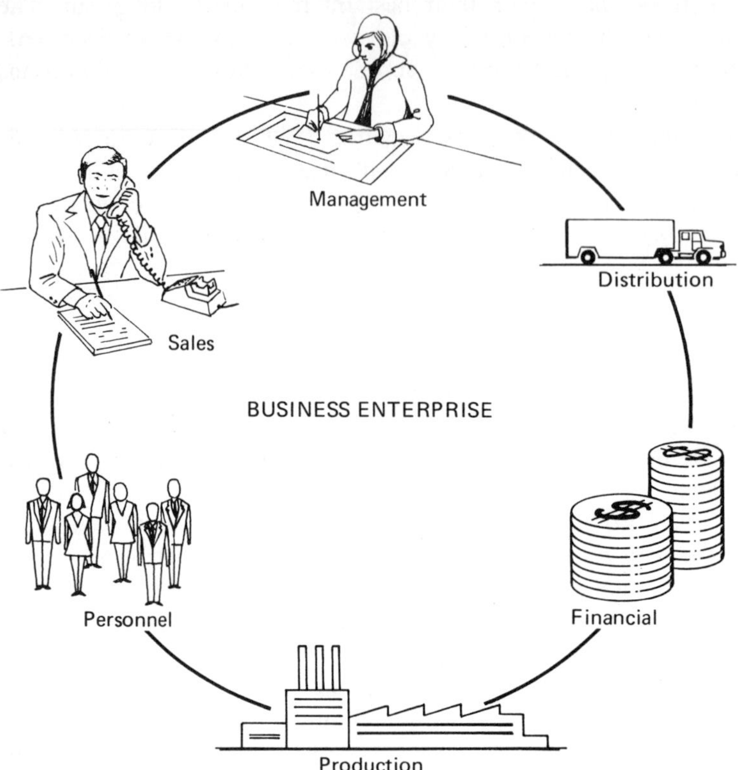


parts or elements. (See Figure 1.4.) Many of these elements themselves are smaller units, called subsystems. Each subsystem performs a specified sub-task that is consonant with the goals of the larger system of which it is a part. For example, a firm's business system might be composed of a communications network, a transportation and delivery system, a personnel department, and a management department.

Designing, specifying, and selecting the elements of a business system falls within the domain of the systems analysts. It is their job to fuse the parts together into a coordinated, cohesive, functioning whole.

## IMPORTANT DEFINITIONS AND TERMS

Systems analysis, like most other disciplines, has developed specialized terms. The most important ones are explained below. Some of the definitions are from the "Vocabulary For Information Processing," American National Standards Institute (ANSI).

**Algorithm:** The logical steps, or sequences of operations, that are followed to arrive at a solution to a problem. The strategy for solving a given problem.

**Constraint:** Limits, or strictures, set upon a problem. The outer limits or rules by which a problem may be solved.

**Data:** A representation of facts, concepts, or instructions in a formalized manner suitable for communication, interpretation, or processing by humans or by automatic means. Facts, figures, names, lists, tables, etc., that are of value to a business enterprise. (Used synonymously with information in this text.)

**Data Bank:** A comprehensive collection of libraries of data. For example, one line of an invoice may form an item; a complete invoice may form a record; a complete set of such records may form a file; the collection of inventory control files may form a library; and the libraries used by an organization are known as its data bank.

**Data Processing:** The restructuring of data to improve its utility to the business enterprise. To change the order, form, or method of storage of data by such operations as sorting, collating, merging, or sequencing.

**Flowchart:** A graphic representation for the definition, analysis, or solution of a problem in which symbols are used to represent operations, data flow, equipment, etc.

**Information:** Meaningful data. Facts, figures, names, etc., which have been organized or processed to increase their value to the business enterprise. (Used synonymously with data in this text.)

**Objective:** A goal toward which effort is directed. An aim or end of action. Objectives are the designated results or outcomes of a system.

**Parameter:** A variable that is given a constant value for a specified procedure or process. The limits or boundaries that define a given problem.

**Policy:** A specific course or method of action, selected from among alternatives and in light of given conditions, to guide and determine present and future decisions. A set of rules upon which present and future decisions are made.

**Procedure:** The course of action taken for the solution of a problem. Those steps that must be followed or traced to solve a problem.

**Source Document:** A form containing information regarding a transaction or activity that is prepared at the time it takes place.

**Subsystem:** A small functioning unit of a larger unit. It is one element of a system and performs a specified sub-task.

**System:** An assembly of methods, procedures, or techniques united by regulated interaction to form an organized whole. An organized collection of people, machines, and methods required to accomplish a set of specific functions.

**System Design:** To define, plan, select, and organize the elements (people, machines, procedures, work flow) that make up a business system.

**System Hardware:** The physical equipment, such as machines, mechanical or electrical devices, used in a system. Generally, computers, adding machines, typewriters, desks, telephones, etc.

**System Software:** The set of computer programs, procedures, and associated documentation concerned with the operation of a data processing system. Includes manuals, diagrams, flowcharts, and language translators used in data processing.

**Systems Analysis:** The methodical study and investigation of a data flow problem, with the view toward improving that flow in terms of maximizing profits, speeding up return of results, and reducing errors.

**Systems Analyst:** An individual charged with the responsibility of assessing business data flow problems, planning, modifying, evaluating, and implementing new or used systems.

**Systems Department:** A group of employees charged with the responsibilities of systems design, implementation, planning, and evaluation.

## EXERCISES

1. List five functions of modern business systems departments.
2. Summarize three advantages of using systems analysis techniques.
3. Describe three limitations of business systems analysis.
4. Define the word "system."
5. List four common examples of systems.
6. Define the word "constraint."
7. Describe the activities involved in systems analysis.
8. Briefly trace the changing emphasis on business systems analysis.
9. How has the need for accountability affected the way in which management views business systems?
10. Explain how human elements complicate the systems analyst's job.
11. How does system hardware differ from system software?
12. Observe the functions of systems in your community. List as many systems as you can, noting whether they are public or private business systems.
13. Select one of the systems listed in exercise 12. Describe its purpose, how it functions, and the constraints under which it operates.
14. Interview teachers and staff on your college campus and describe the system used for registering students.
15. Interview the personnel in your campus bookstore and describe the system used for ordering books.

# Chapter 2

# Structure of Business Organizations

By definition, a business organization is an administrative unit designed to coordinate the elements of the firm into an integrated, coherent, functioning whole. The administrative unit is usually structured, or organized, in one of several common forms.

Before a systems analyst approaches the task of developing or redesigning a business system, he or she must understand how the firm is organized and what are the formal and informal chains of command and action. The analyst must be able to trace communications and data flow through this structure, and must determine the points in the organization where data originates and where it is output.

This chapter describes the structure and anatomy of the typical business firm in terms of both its classical, or formal, organization and its pragmatic, or informal, arrangement. It also discusses the common patterns of communication and data origination and flow.

## THE FORMAL ORGANIZATION

Organizations are composed of individuals who perform specified roles coordinated to enable the company to carry out its goals. Virtually every firm has some type of official structure or formally described organization.

It has some official plan that clearly shows the chain of command and the lines of authority and responsibility. It has some formalized set of rules that defines the various functions and duties incurred by each level of authority and prescribes the relationships among the different roles. It delineates the line of authority and spells out officially, and clearly, to whom an individual is responsible, and the employees for whom he or she is responsible.

Most business firms are structured according to one of the following three types of formal organizations:

CLASSICAL LINE ORGANIZATION. A typical classical organization resembles a pyramid. At the top is the leader—chairman of the board, president, chief—with the ultimate decision-making authority. Below the leader, on different levels, are the other individuals who make up the firm.

On the level directly below the leader are the relatively few persons who are next in authority and to whom have been delegated lesser or different responsibilities and decision-making functions. Below them are the other layers of the official hierarchy, each involving a greater number of individuals with lesser responsibility and authority than the one above.

Within this classical line organization exist clear lines of authority, moving from the top down. Each individual is responsible to someone in authority above him or her. Conversely, each individual in authority has persons for whom he or she is directly responsible.

These relationships and lines of authority are often illustrated with a visual, graphic device called an organization chart. Figure 2.1 shows the lines of authority and chain of command in a firm organized in the classical line pattern. If individuals are not connected by lines, it is assumed that no authority or responsibility exists between them.

The classical line organization is a simple, clear structure, but it does contain certain weaknesses. Primarily, to receive any guidance an individual needing advice and counsel must follow the line of authority shown on the organization chart. He or she cannot approach other organization members not related by a direct line of authority, since no official relationship has been established.

Because of this weakness, the line organization is not widely used in modern business or industry. The military is probably the remaining classic example of this type of organization.

LINE AND STAFF ORGANIZATION. A modified form of business organization is the line and staff structure shown in Figure 2.2. (Line positions are indicated by solid lines; staff positions by dotted lines.) This type combines the usual line positions, representing the formal chain of command, with supplemental staff positions.

Staff positions refer to those employees and departments that provide a consulting or advisory service to other departments in a firm. Usually such departments are not directly involved in producing the goods or services that are the firm's major source of profits. Examples would be personnel, accounting, legal, research, or data processing departments.

The staff member's role is to provide consultation, advice, and support

**Figure 2.1.** Classical line organization

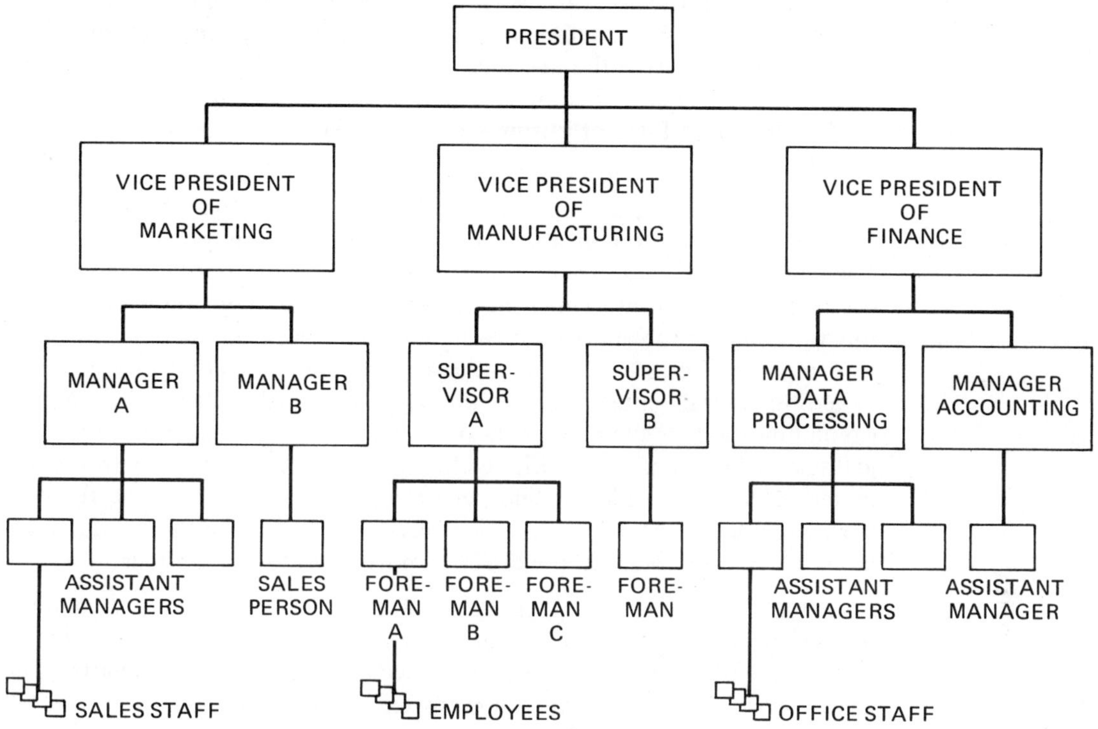

PRESIDENT
VICE PRESIDENT OF MARKETING
VICE PRESIDENT OF MANUFACTURING
VICE PRESIDENT OF FINANCE
MANAGER A
MANAGER B
SUPER-VISOR A
SUPER-VISOR B
MANAGER DATA PROCESSING
MANAGER ACCOUNTING
ASSISTANT MANAGERS
SALES PERSON
FORE-MAN A
FORE-MAN B
FORE-MAN C
FORE-MAN
ASSISTANT MANAGERS
ASSISTANT MANAGER
SALES STAFF
EMPLOYEES
OFFICE STAFF

**Figure 2.2.** Line and staff organization

PRESIDENT
VICE PRESIDENT MARKETING
VICE PRESIDENT PERSONNEL
VICE PRESIDENT MANUFACTURING
MANAGER
MANAGER
MANAGER
MANAGER
MANAGER
MANAGER

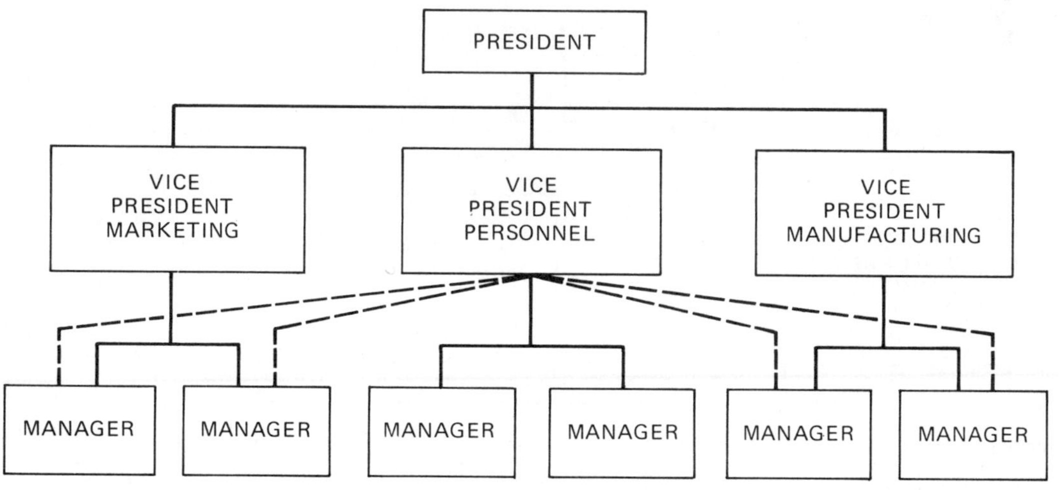

to the line officers in areas outside their expertise. The actual responsibility for the end performance, however, remains with the line officer.

Provisions for consultation and staff support, in addition to the clear lines of authority, give the line and staff organization an important advantage over the classical line structure. For this reason it has become the most widely used form of business organization.

COMMITTEE ORGANIZATION. The third form of business structure is the committee organization, shown in Figure 2.3. In this arrangement, a position is assigned to a group of individuals rather than to a single person. The group shares the responsibilities delegated to that position and makes decisions by consensus. Thus, the talents of several people are involved in solving a single problem or in making a difficult decision.

A main weakness of the committee organization is the difficulty involved in getting several people to act cohesively and to make decisions promptly. Communications among members of a committee can also pose problems, and it may be difficult for an individual to function effectively with a committee rather than with a single person as the "boss." For these reasons, the committee setup is not a widely used form of business structure, although it is often found in political and government organizations.

DEPARTMENTAL ORGANIZATION CHARTS. The departments of a business enterprise vary in structure and complexity, depending on the activities they perform. These variations are reflected in organization charts. The charts shown below illustrate some typical departments.

Figure 2.4 is the portion of an organization chart that illustrates the manufacturing department of a firm with three plants. At the top is the vice president of manufacturing, responsible for all activities in the department and reporting directly to the company president. Below the vice president, in line positions, are three plant managers, each in charge of one plant.

Each plant manager has several supervisors in charge of such areas as

**Figure 2.3.** Committee organization

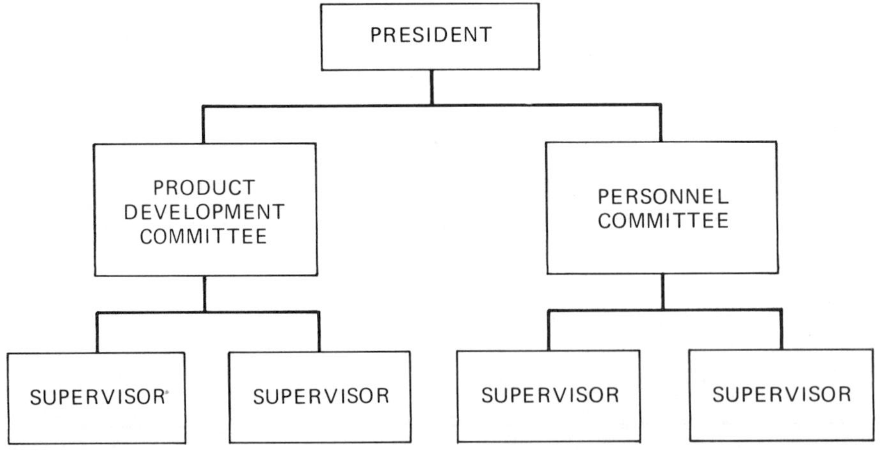

**Figure 2.4.** Manufacturing department

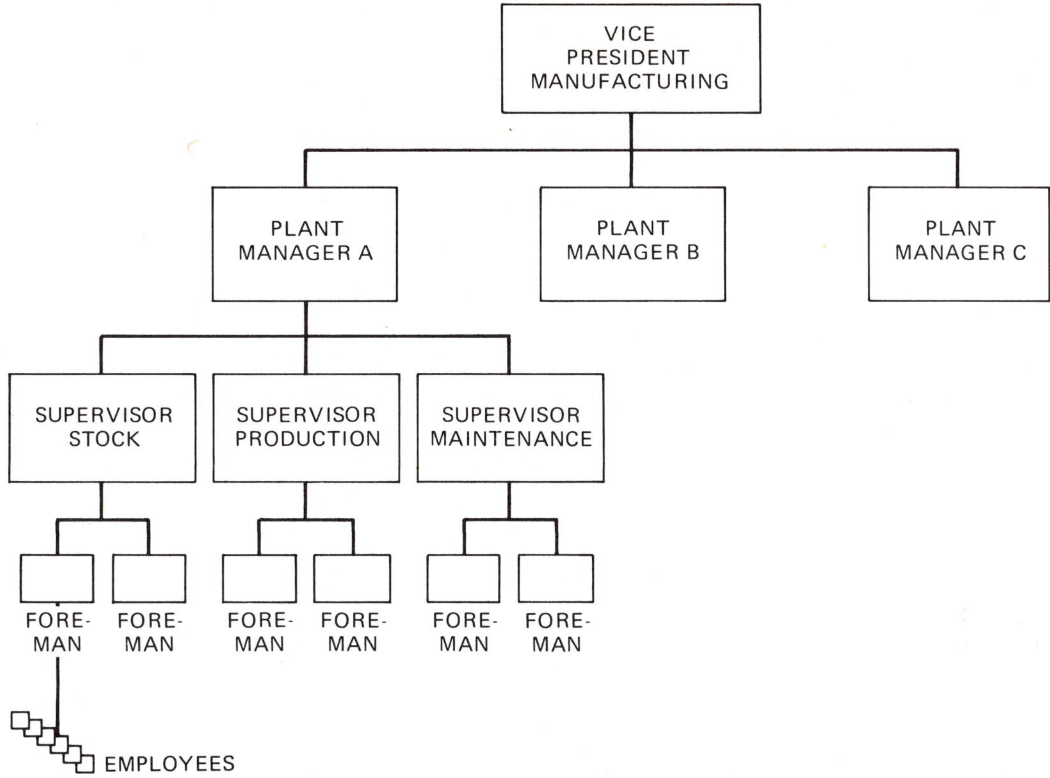


stockkeeping, production, and maintenance. Below each supervisor are several foremen. Each foreman reports to an immediate supervisor and, in turn, is responsible for the activities of several employees.

Figure 2.5 illustrates the marketing and sales division of an organization. The division is composed of three sales departments—wholesale, retail, and export. The vice president of marketing holds the decision-making authority and responsibility for the division. Lesser responsibility and authority are delegated to three managers, each of whom runs a department.

The wholesale and retail sales departments are divided into smaller territories, each headed by an assistant sales manager. The export sales department has a different structure, with the entire responsibility resting on the department manager.

Figure 2.6 is a typical organization chart for an accounting division. At the top of the chain of command is the vice president of finance, who oversees all operations within the division. Below that position are several managers in charge of data processing, payroll, accounts receivable, accounts payable, cost accounting, and other areas. Within each of the departments are several supervisors and numerous employees.

These examples illustrate only the line positions in the organizations. Staff relationships are not shown. As a practical matter, each vice president

**Figure 2.5.** Marketing and sales division

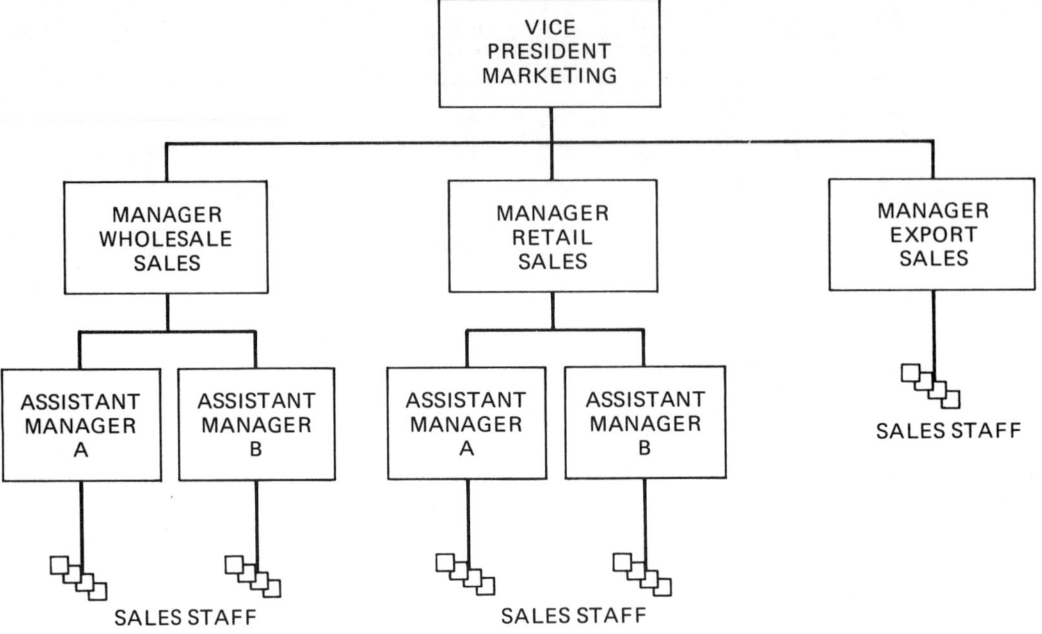


**Figure 2.6.** Accounting division

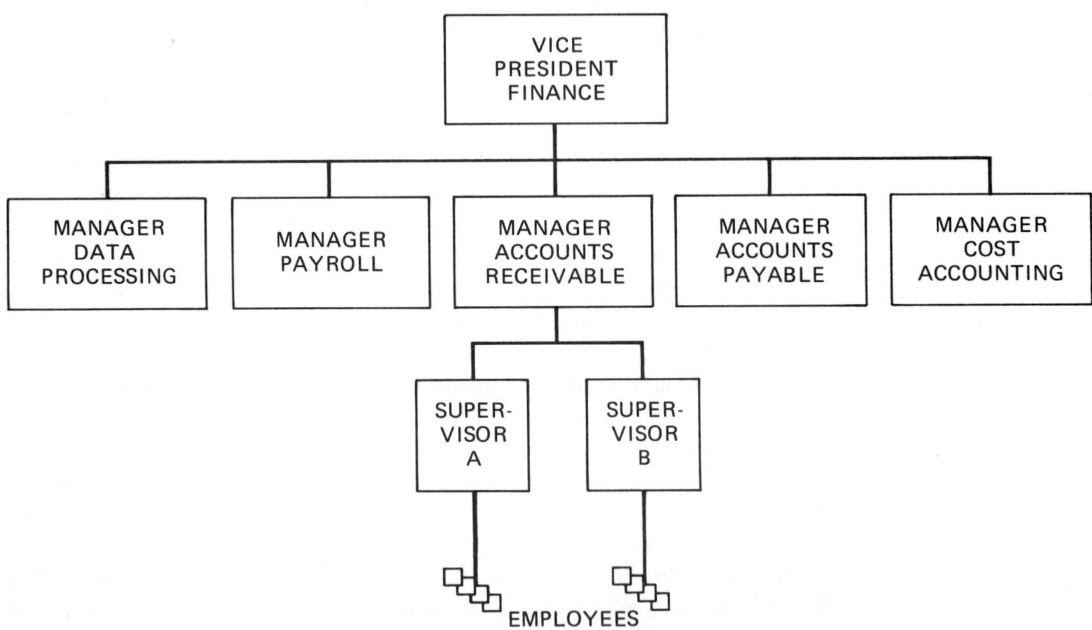

has staff personnel available. This allows him or her to draw upon the expertise of others for consultation and advice when necessary, and still maintain direct line authority over subordinates. Thus, the marketing vice president may serve a staff function for manufacturing. This assures that the manufacturing department produces goods that meet the needs of the marketing division. Similarly, the accounting department head may have a staff relationship with the manufacturing and marketing departments, handling their payroll, costing, and other accounting procedures.

## INFORMAL ORGANIZATION

The lines and pictures on the organization chart depict the official lines of authority and relationships as they should be, or are expected to be, in a business firm. But in the everyday working environment, many other factors and relationships intrude and influence this formal description. These unstated, pragmatic roles and patterns that govern the actual activities of a business make up its "informal" organization.

The informal structure of a business is difficult to describe. It does not appear on the organization chart; it varies with the firm; and it may be different from, and in conflict with, the firm's stated goals. Yet it is present in nearly every business enterprise, exerting pressures and guidance that may be equal to, or may exceed, those of the formal organization.

The informal organization is built around key individuals who have an unofficial and indirect, but important, effect on the relationships and decision-making responsibilities of others in the company. They are the people to whom other employees turn for help during day-to-day working experiences, rather than to the ones actually assigned on the official organization chart. They may be someone else's assistant, a supervisor or secretary, or perhaps a peer member not in a line or staff relationship.

The organization chart in Figure 2.7 shows the relationship between the informal and formal structures of a firm. An assistant manager in the public relations department turns to the manager in the marketing department for advice and help, instead of discussing a situation with the manager officially above him in the line of authority. The vice president of personnel relies on a member of the office staff for knowledge of employee attitudes and reaction, and also tends to discuss problems with the marketing manager before making decisions.

Identifying and evaluating the elements and relationships in a firm's informal organization is an important, but difficult, part of the systems analyst's job. The influence and reactions of the leaders of the informal organization toward a new or modified system should not be underestimated or ignored. Other employees may often observe and copy the informal leaders' reactions to the situation, rather than following the recommendations of the official authorities. Soliciting the help of these informal leaders will go a long way toward insuring the success of the efforts of the systems analyst.

**Figure 2.7.** Formal and informal organizations

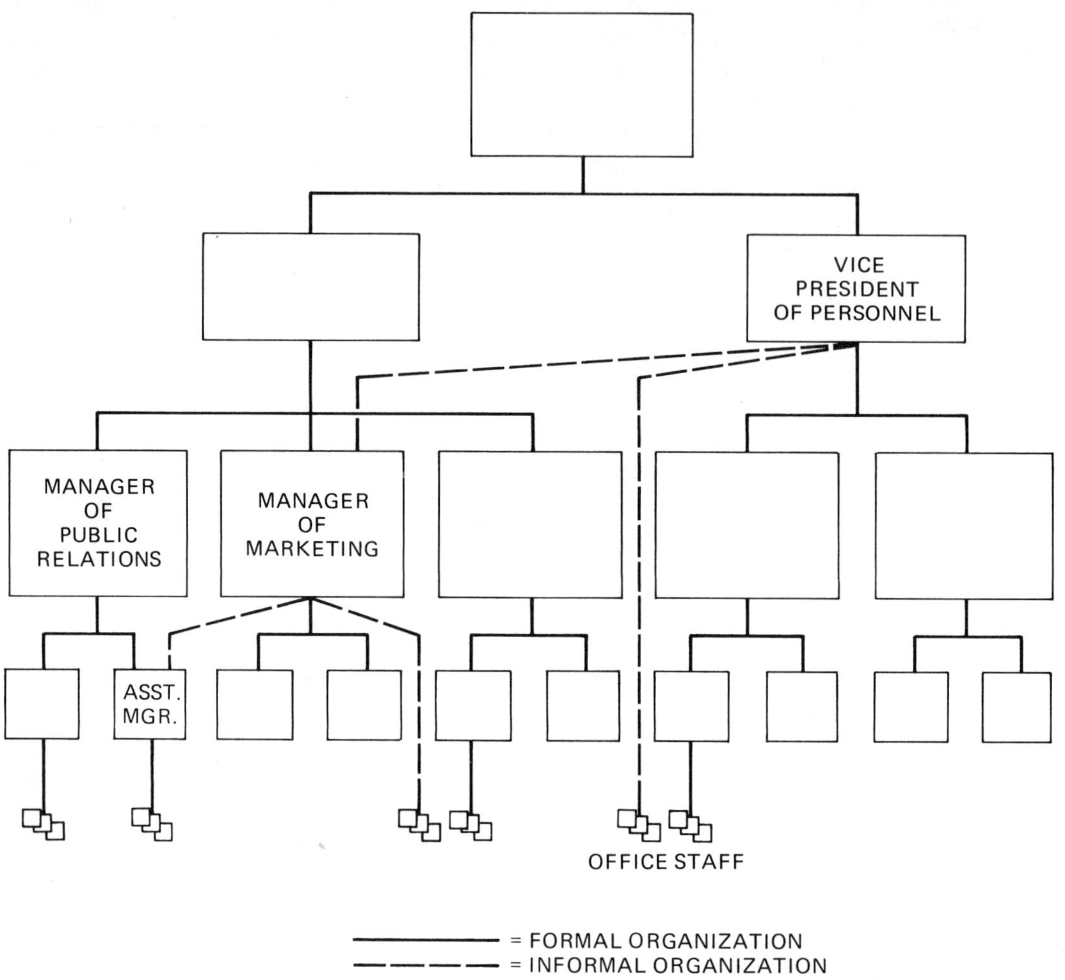


## PEER GROUPS AND STATUS

Other factors that affect the official relationships of the formal organizational structure include peer group influence and status symbols. While a thorough discussion of these sociological, psychological, and motivational factors are beyond the scope of this book, the astute systems analyst will be aware of the part these elements play in influencing employee behavior, expectations, and reactions.

Peer group influence often carries more weight than official statements from superiors. It is very effective in encouraging fellow employees to adhere to the standards in the informal organization rather than to those of the official structure. An individual who does not conform to these patterns risks being pressured or ostracized by his or her peers.

The status symbols and non-monetary rewards that come with a position or role also influence relationships, reactions, and motivations. A status symbol—a name on the office door, a reserved parking space, a private secretary—may bring greater authority and respect, and more affect the patterns of the informal relationships than would an increase in salary.

The effects of peer groups and status symbols can make it more difficult for the systems analyst to evaluate the formal structures of an organization accurately and objectively.

## COMMUNICATIONS AND THE ORGANIZATION

Communications, both verbal and written, are vital elements in the daily life of a business enterprise. Identifying and evaluating the activities and work stations in a firm that generates data is one of the basic jobs of the systems analyst. He or she must be able to trace its flow and transformation throughout the organizational structure in order to understand and evaluate a data flow system.

Tracing communications flow is facilitated first by defining the kind of communication that is taking place. One way is to classify it according to destination—is it internal or external? Is it directed to another person or department within the same firm? Or to another company or individual outside the firm?

*External communications* are those that take place between an organization and the other entities with which it interacts. Examples are communications that flow between a firm and its customers, the government, stockholders, unions, and others. (See Figure 2.8.) Documents include purchase orders, bills of lading, sales slips, customer invoices, payments to vendors, tax reports, and more.

*Internal communications* are letters, memos, and other correspondence that originate and terminate within the same organization. They play an important part in the functioning of a business enterprise, and include such items as stock records, work reports, expense account records, and check registers.

These internal communications can be further classified by direction—is it horizontal or vertical? *Horizontal communications* include the transfer of data and information between individuals at the same level on the organization chart. (See Figure 2.9.) Examples are one clerk typist teaching another to prepare an invoice; a report prepared by one office for another on the same level; and the transfer of sales orders from the sales department to accounting.

*Vertical communications* is the transfer of information between individuals on different levels of the organization chart. (See Figure 2.10.) Examples include a request for information from a vice president of manufacturing to a line foreman, or a shipping clerk forwarding a report to his or her supervisor.

**Figure 2.8.** External communications

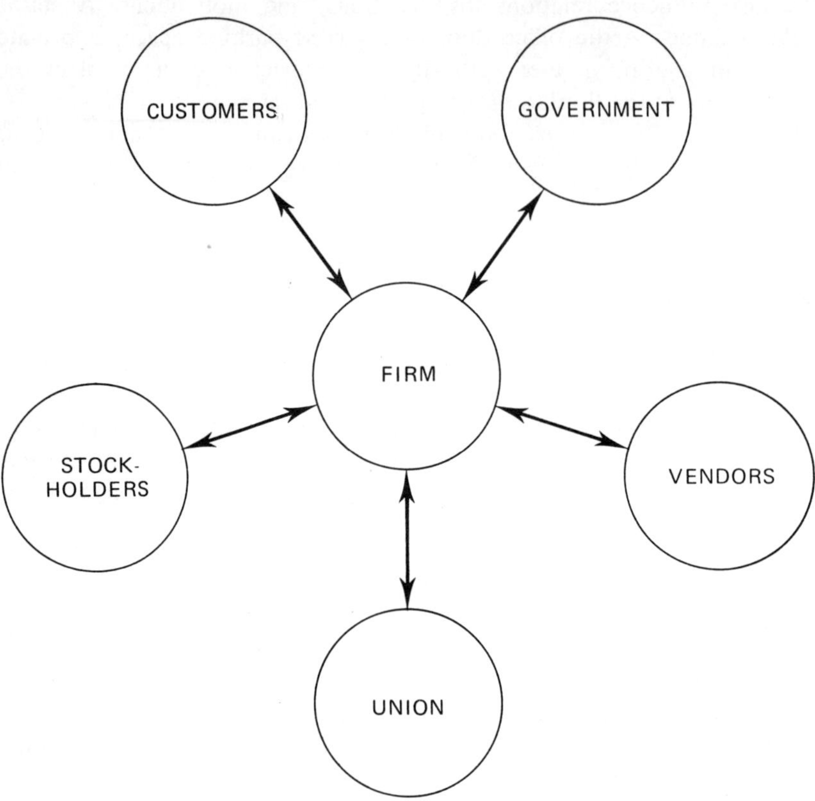


**Figure 2.9.** Horizontal data communications

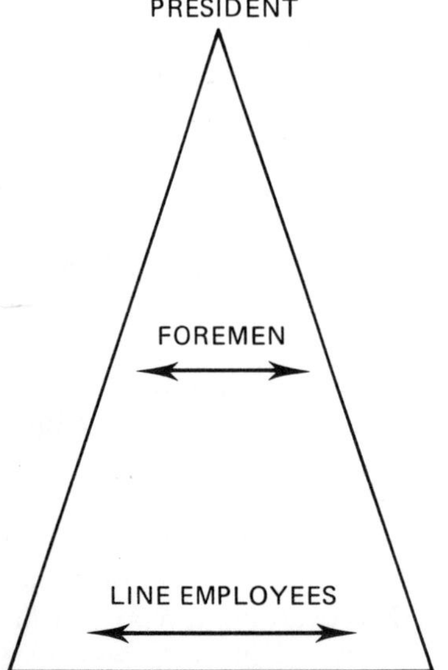


**Figure 2.10.** Vertical data communications

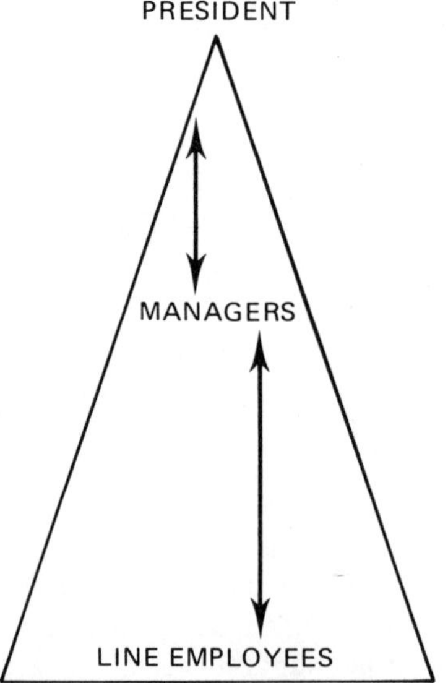

## MAJOR DATA PRODUCING ACTIVITIES

A study of business organizations reveals that the vast amount of paperwork and data generated by a firm originates from nine major activities.[1] These activities are found, to one degree or another, in most business organizations. Each activity involves the generation, processing, checking, transmittal, and filing of many forms, records, or documents.

1. Purchasing activity (Figure 2.11). The purchasing activity is concerned with buying the necessary machines, devices, services, equipment, and supplies necessary to carry out the operations of the business.

**Figure 2.11.** Purchasing activity

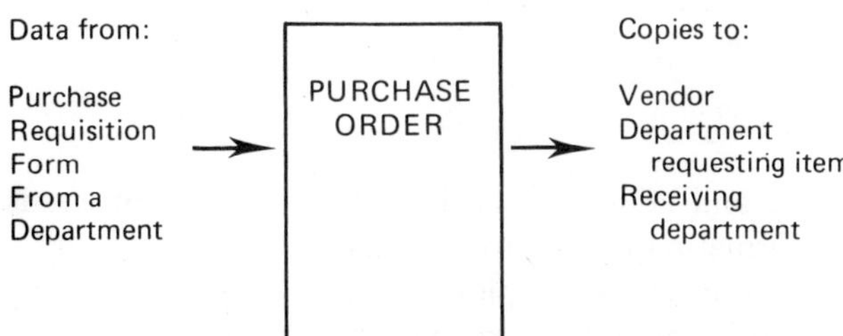


The purchase order is the major document that is generated. It contains the data necessary for ordering supplies needed by departments in the firm. It is initiated when a department prepares a purchase requisition form requesting specific goods, and sends it to the purchasing department. A purchase order is prepared showing such data as type of merchandise, quantity, description, and shipping instructions. Copies are sent to the vendor from whom the goods will be bought, to the department that made the initial request, and to the receiving department.

2. Receiving activity (Figure 2.12). The primary document for this operation is the receiving record. After the goods have been received, a receiving record is prepared. It may be a copy of the purchase order or a special form. It shows such data as the physical count of the items received,

**Figure 2.12.** Receiving activity

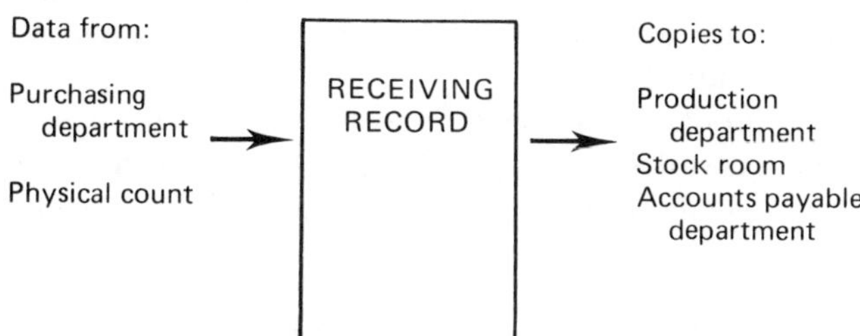


[1] Adapted from "Forms for the Nine Key Operations of Business," Moore Business Forms, Inc.

charges, and descriptions. Copies of the receiving record are sent to the production department, to the stock room, and to accounts payable.

3. STOCKKEEPING ACTIVITY (Figure 2.13). In this activity a record of all goods in stock is maintained. The stock record is the major document in this operation. It is a ledger card on which is kept a running balance of the goods in stock. The record is updated to show when new goods are

**Figure 2.13.** Stockkeeping activity

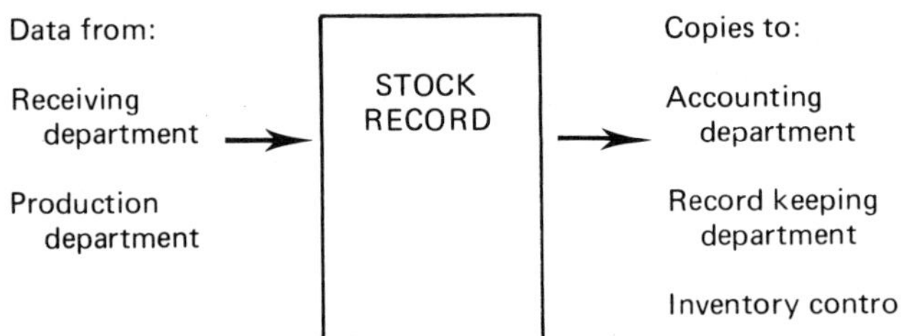


manufactured or purchased and added to the inventory, and when goods are issued or sold. This information is used to generate purchase requisition orders for items running low in stock. Copies are sent to the accounting department, inventory control, and record keeping departments.

4. PRODUCTION AND MANUFACTURING ACTIVITY (Figure 2.14). This activity centers around the manufacture, production, or assembly of goods. The major document used is the production order. It contains information

**Figure 2.14.** Production activity

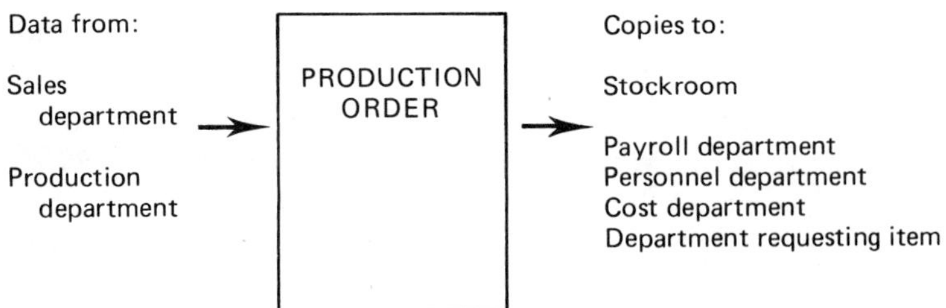


from the sales or production department regarding the manufacture of a given lot of goods. It shows such data as salesperson, customer, part number and description, standards, and scheduling.

The information on the production order goes to the stockroom, payroll and personnel departments, cost department, and such. A copy of the order sometimes travels with the goods through the entire manufacturing operation.

5. SELLING ACTIVITY (Figure 2.15). This activity is involved with the sale or marketing of the goods. The sales slip or tag is the major document in use. It shows such things as price, customer, department number, item number, terms, and date. In retail sales, orders come in via personal contact,

**Figure 2.15.** Selling activity

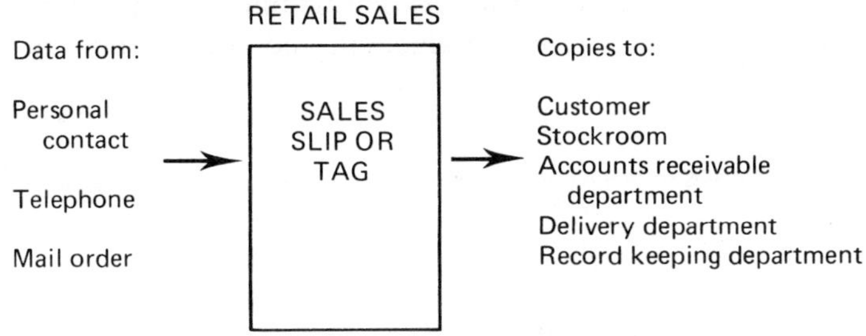


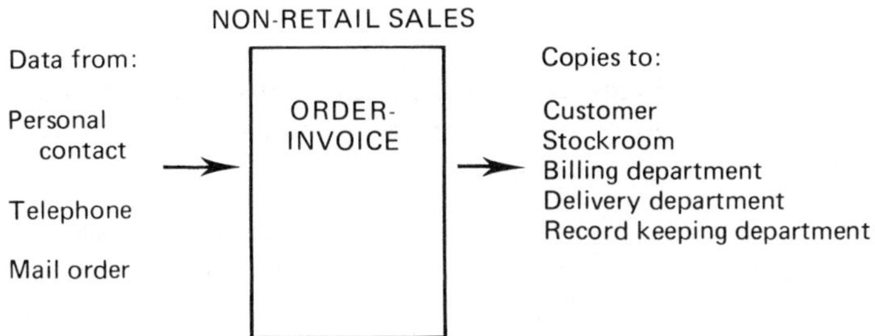


mail order, or telephone. A sales slip is prepared or a sales tag is removed from the goods. Copies of sales forms go to the customer, stock, accounts receivable, delivery, and record keeping departments.

Non-retail activities, such as wholesale or jobbing, use order invoices. They contain such data as price, customer's name, delivery dates, salesman's name, item number, and special instructions. Copies go to the stockkeeping, billing, delivery, and record keeping departments, and to the customer.

6. DELIVERY ACTIVITY (Figure 2.16). In this operation manufactured goods, or items from stock, are delivered to the customer, generating the shipping memo. The memo contains information received from the stock-keeping, production, sales, or purchasing departments. It includes customer's name, purchase order number, dates, shipping instructions, weight, type of merchandise, and COD charges. Copies of the shipping memo go to the delivery company, truck driver, and customer, or are shipped with the goods.

**Figure 2.16.** Delivery activity

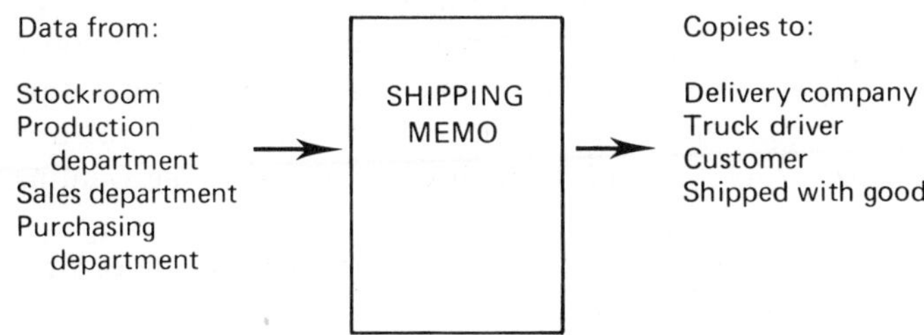

7. BILLING ACTIVITY (Figure 2.17). Invoices are prepared for goods once they have been sold and delivered to customers. Information for the invoices comes from the delivery, sales, production, and stockkeeping departments, and includes such items as names, prices, terms, and part numbers. Copies of the invoices go to the customer, accounts receivable, and accounting and record keeping departments.

**Figure 2.17.** Billing activity

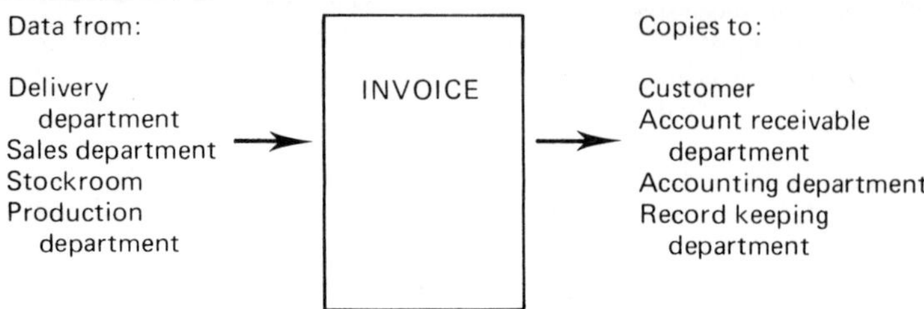


8. COLLECTING ACTIVITY (Figure 2.18). This operation consists of maintaining the accounts receivable records, sending collection notices, and recording payments. The primary document used is the statement. It is sent periodically to the customer, and notes the number of invoices, their dates and amounts. Sometimes a payment stub, or money receipt, is returned by the customer to facilitate posting payments. Other copies go to the accounts receivable department for follow-up and to the accounting department.

**Figure 2.18.** Collecting activity

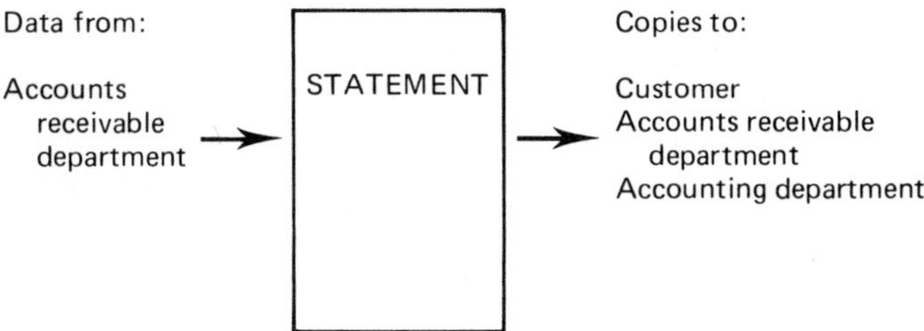


9. DISBURSING ACTIVITY (Figure 2.19). The payroll and payments to vendors are handled in this operation. The compensation record is the primary document used for preparing the payroll. It contains all the information necessary for computing employees' paychecks—rate, hours worked, overtime, deductions, exemptions, and tax withholding. Information comes from the personnel department, the department where the employee works, from time cards, expense accounts, and others. The information on this record is used to generate the employee's paycheck, federal and state tax returns, entries to the check register, and other accounting records.

The voucher check is the primary document used to pay vendors for

**Figure 2.19.** Disbursing activity

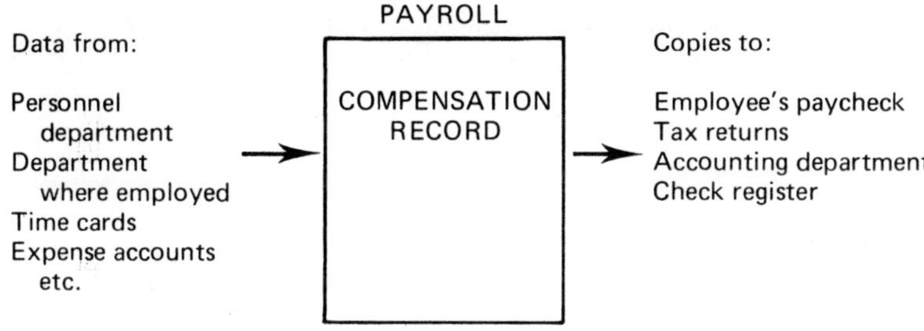


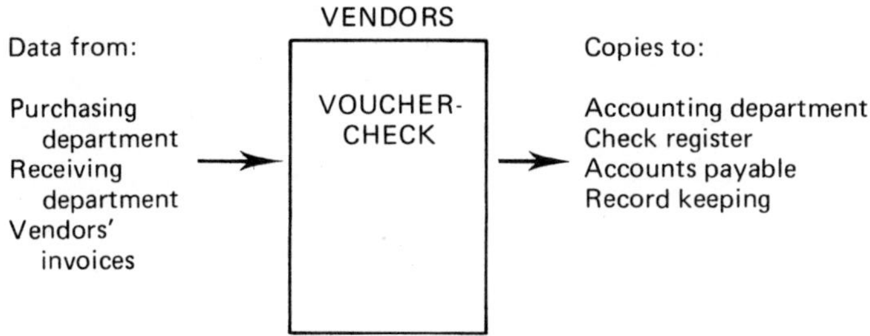


services rendered or goods purchased. Requests for payment of a bill come from the purchasing department, receiving department, or vendors' invoices. Copies of the check voucher or data related to the paid invoice go to the accounting department, check register, accounts payable, and record keeping activities.

## EXERCISES

1. Describe at least two formal business organizational structures.
2. Describe the function of departmental organization charts.
3. Define the term "informal organization."
4. How does the formal organization differ from the informal organization?
5. What effects do peer groups and status symbols have upon the formal organization?
6. Contrast the differences between external and internal communications.
7. Give several examples of horizontal communications in an organization.
8. List at least six major data producing activities in a typical business organization.
9. Describe the record typically generated by the purchasing activity.
10. Describe the records typically generated by the disbursing activity.
11. Draw a formal organization chart showing the relationships of personnel on your college campus.
12. Interview an employee of a business firm with whom you are familiar, and discuss the informal structure. Draw an organization chart showing these informal relationships.

13. Gather six miscellaneous documents produced by an organization. Determine which major activity generated each document.
14. Write a brief essay describing the advantages and limitations of a committee structure in solving problems and making decisions.

# Chapter 3

# Basic Business Systems

A system can be defined as an organized collection of people, machines, and methods required to accomplish a set of specific functions. A business system is designed to perform the function of generating and processing the data necessary for the successful operation of the firm—efficiently, accurately, and at the lowest cost.

Since firms differ in their major services, products, and goals, they require different kinds of data and processing procedures. The elements of the business system must be tailored to meet the specific needs of the individual company.

Firms that manufacture goods generally build their systems around the order. Banks and lending institutions focus on the account. Insurance companies center their systems on the policy, and listings are the focal point of real estate firms. Printers rely on the job ticket; airlines, hotels, and motels on the reservation; and stock brokers are dependent on the buy and sell order. The business systems that serve these varied needs have different focal points, but they contain elements common to all.

### Order Processing System

This is one of the most widely used business systems. All data generation, processing, and movement centers around the order and its progress through the firm. This system is used by companies that manufacture goods or products, wholesalers and retailers, printers, publishers, and other types of businesses in which the main activity is the fabrication or marketing of a good.

The steps in the order processing system (shown in Figure 3.1) are described below:

**Figure 3.1.** Order processing system.

**Figure 3.2.** Order form

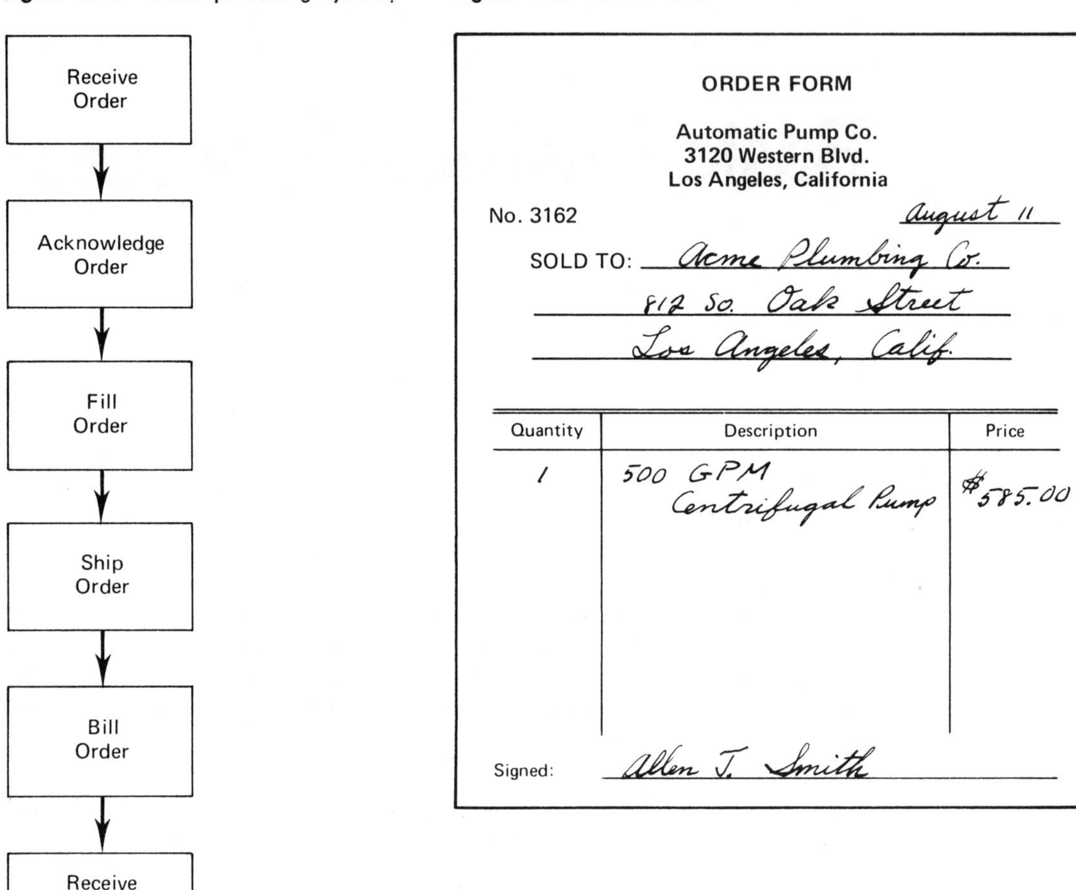


ORDER FORM

Automatic Pump Co.
3120 Western Blvd.
Los Angeles, California

No. 3162 August 11

SOLD TO: Acme Plumbing Co.
812 So. Oak Street
Los Angeles, Calif.

| Quantity | Description | Price |
|---|---|---|
| 1 | 500 GPM Centrifugal Pump | $585.00 |

Signed: Allen T. Smith

1. Receive order. The first step is receipt of the order by person, telephone, mail, or messenger service. The indicated specifications are recorded on some type of ledger or form, such as a job ticket, work order, or sales order (Figure 3.2).

2. Acknowledge order. After the order is received and recorded, some form of acknowledgement is usually sent to the customer. This may consist of a postcard acknowledging receipt of the order and showing the anticipated delivery date; a copy of the order form; or a written estimate showing plans, layouts, and cost quotations. (See Figure 3.3.)

**Figure 3.3.** Form used to acknowledge order

Automatic Pump Co.
3120 Western Blvd.
Los Angeles, California

SOLD TO: Acme Plumbing Co.

This is to acknowledge receipt of your order of August 11.

Thank you for your patronage.

Roy E. Howard
Order Department

3. FILL ORDER. This step encompasses all the activities related to producing or acquiring the ordered goods. Materials are ordered from suppliers or manufacturers, or removed from stock, and manufacturing processes are performed. The fill order generates data regarding expenditure of funds for goods and services from outside sources, changes in inventory status, sales activities, and related labor costs. (See Figure 3.4.)

4. SHIP ORDER. The ordered items have now been picked from stock, or manufactured, and are ready for shipment to the customer. Bills of lading, shipping memos, labels, and such are prepared to accompany the goods. (Some examples are shown in Figure 3.5.)

5. BILL ORDER. After the order has been filled and the goods delivered, the charges are billed to the customer. This involves the preparation of an invoice showing the costs incurred during manufacture, removal from inventory, shipping, and others. In addition, a statement summarizing the charges on a group of invoices may be mailed to the customer. (Some examples are shown in Figure 3.6.) Copies of invoices are sent to the accounting department for payment and collection.

6. RECEIVE AND POST PAYMENT. The last step in the order system is the receipt of payment. This data must be deducted from the balance due on the customer's account and posted to the company's accounting books. This final step completes the order processing cycle.

### Inquiry System

This is a major type of system used by hotels, airlines, banks, credit companies, insurance companies, and other firms that maintain large information or physical inventories. Its focal point is the data base, and data processing activities are related to changes in its status.

The inquiry system consists of a central file (data base) and a network

**Figure 3.4.** Data generated when filling order

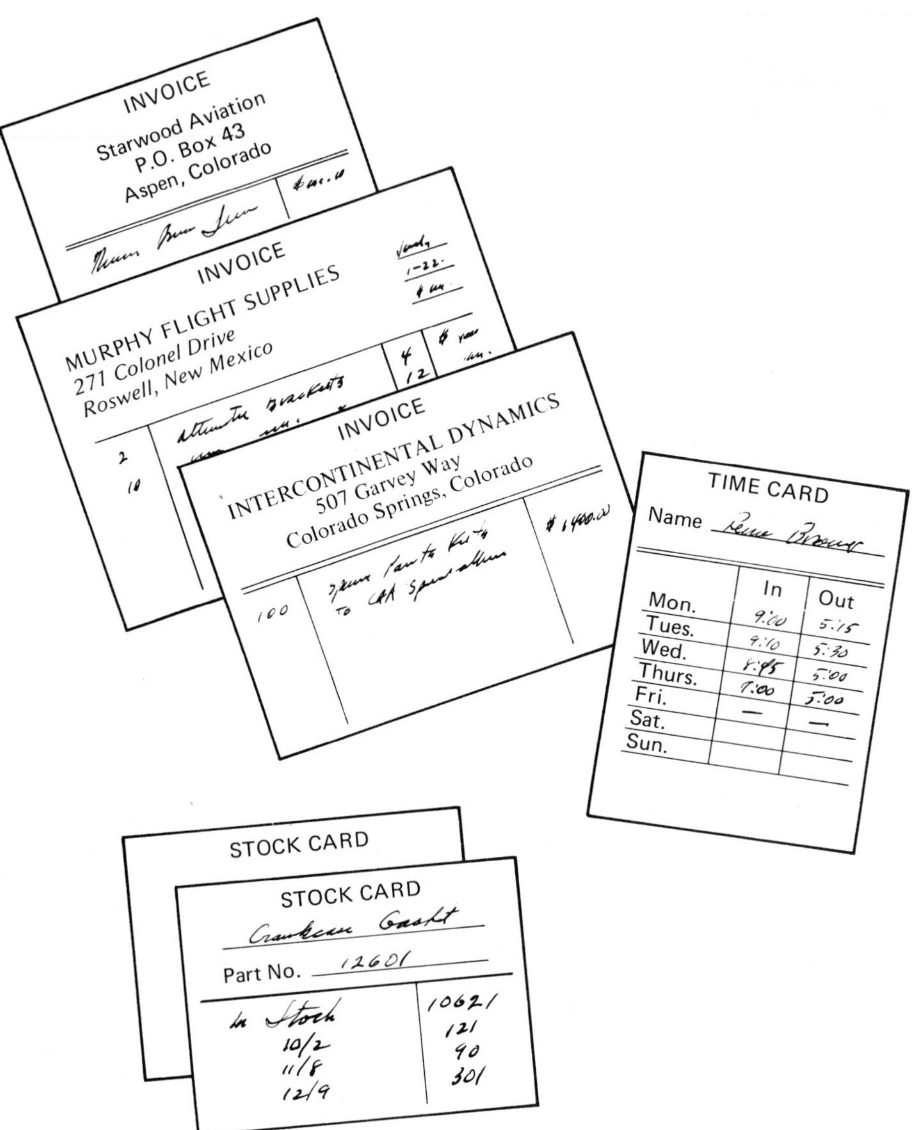


of query stations (Figure 3.7). The data base contains facts and other data related to the physical inventory or body of information. A data base for hotels and airlines would contain information regarding such things as available reservations, connecting flights, time schedules, and dates. A company that stocks a large number of parts or goods would record information such as part numbers, prices, and quantities. Banks and credit departments would record information on customer accounts, and insurance companies would store actuary data and general policy information.

**Figure 3.5.** Data generated when shipping order

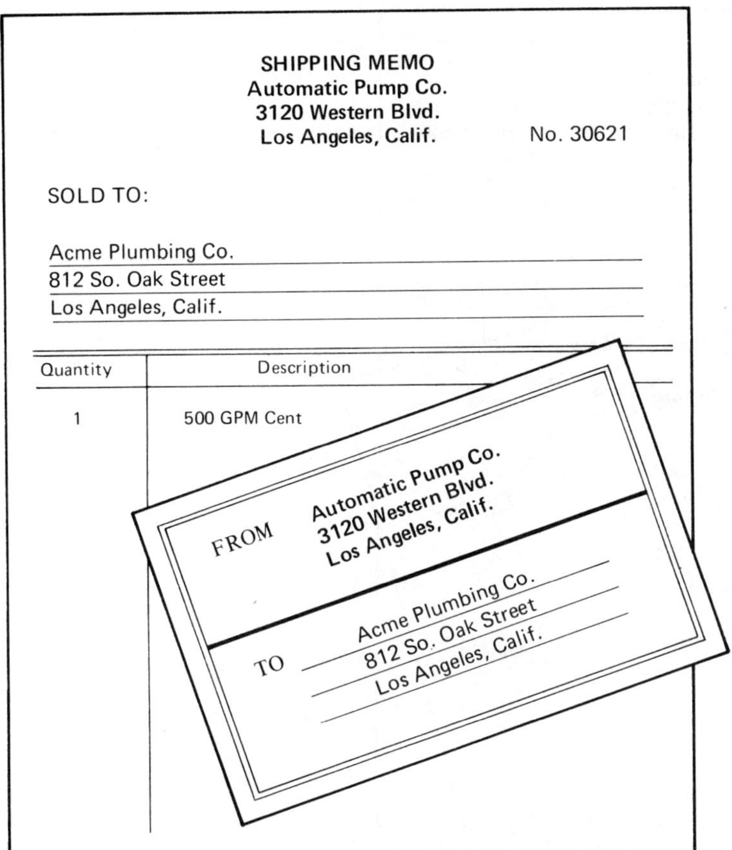

SHIPPING MEMO
Automatic Pump Co.
3120 Western Blvd.
Los Angeles, Calif.

No. 30621

SOLD TO:

Acme Plumbing Co.
812 So. Oak Street
Los Angeles, Calif.

| Quantity | Description |
|---|---|
| 1 | 500 GPM Cent |



The query stations, or terminals, are the locations from which users access the data base. These stations may be in the same room as the data base or at a remote location. Communications between the data base and the query stations are provided by teleprocessing networks, telephone lines, messengers, computer terminals, Teletype terminals, and others.

The purpose of the inquiry system is to make a central body of information available to many users. This data must be current and must reflect changes in the status of the data base or inventory as they occur. As parts are removed from stock, the quantity recorded in the data base is altered to reflect the change. When a withdrawal is made, the balance on the customer's account is adjusted. If new actuary data becomes available, it is entered in the central file. If new flights are scheduled, data showing the time schedule, costs, or seating arrangements is placed in the data base to keep its information current.

Some data base arrangements allow the system to be updated from remote terminals; others from a central location only. An inquiry might go

Figure 3.6. Data generated when billing order

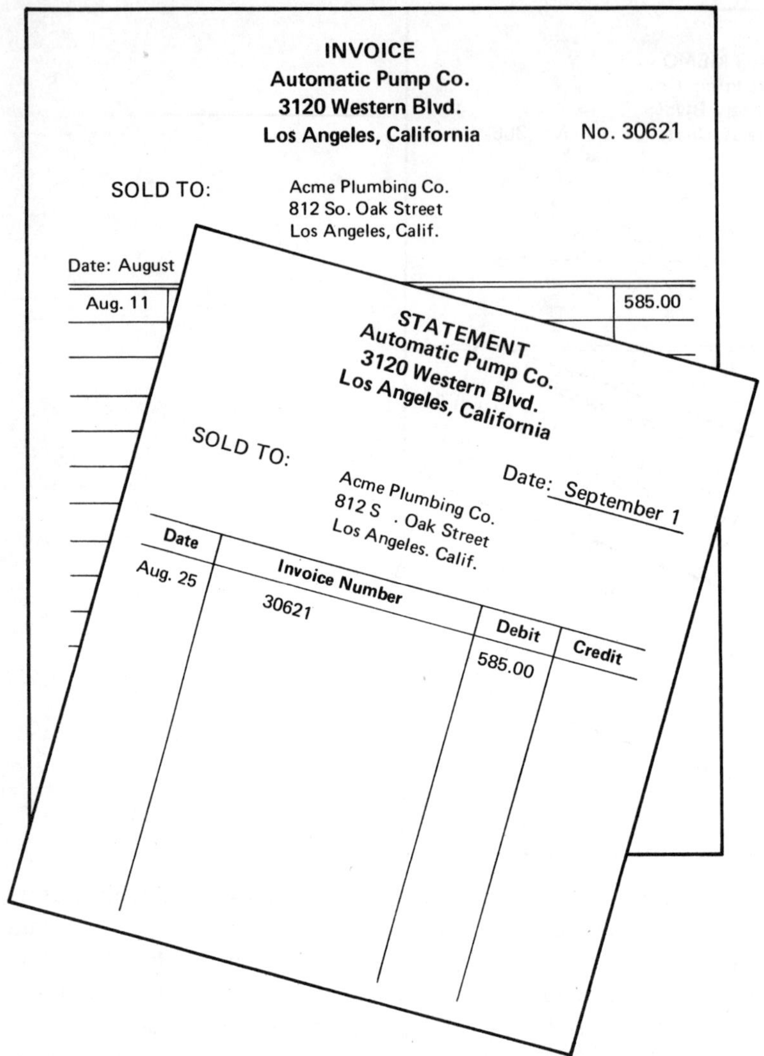

INVOICE
Automatic Pump Co.
3120 Western Blvd.
Los Angeles, California

No. 30621

SOLD TO: Acme Plumbing Co.
812 So. Oak Street
Los Angeles, Calif.

Date: August

Aug. 11 ... 585.00

STATEMENT
Automatic Pump Co.
3120 Western Blvd.
Los Angeles, California

SOLD TO: Acme Plumbing Co.
812 S . Oak Street
Los Angeles. Calif.

Date: September 1

| Date | Invoice Number | Debit | Credit |
|---|---|---|---|
| Aug. 25 | 30621 | 585.00 | |

directly to the data base from a remote terminal, or it might be received by an operator at the data base site who queries the file and relays the information back to the terminal.

Most inquiry systems involve the following steps (shown in Figure 3.8):

1. Establish the basic file. A master file is set up listing the current information (parts available in stock, seats on a flight, room accommodations, credit limits). Data relevant to each item available is also included. (See Figure 3.9.) The file may be recorded on one of several media (shown in Figure 3.10), including computer storage devices, tub

**Figure 3.7.** Inquiry system

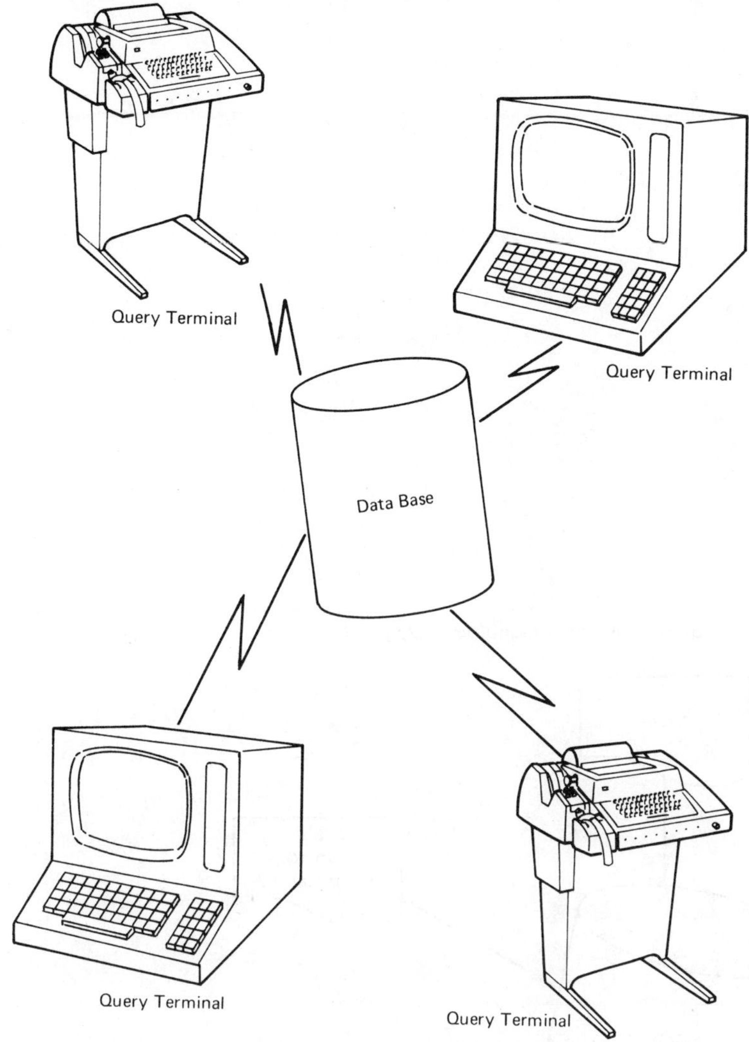


files, and card index systems. The computer is probably the most widely used system today for large files.

2. PROCESS QUERIES. Inquiries are received and processed through a network of stations from on-site and remote locations. For example, reservation clerks stationed at airports and branch offices query the central data base via computer terminals for the current information on available seats and schedules. Sales people in department stores request approval, via teleprocessing, of charge customers. Employees in a manufacturing plant place queries to the central inventory system through a Teletype terminal to

**Figure 3.8.** Inventory system

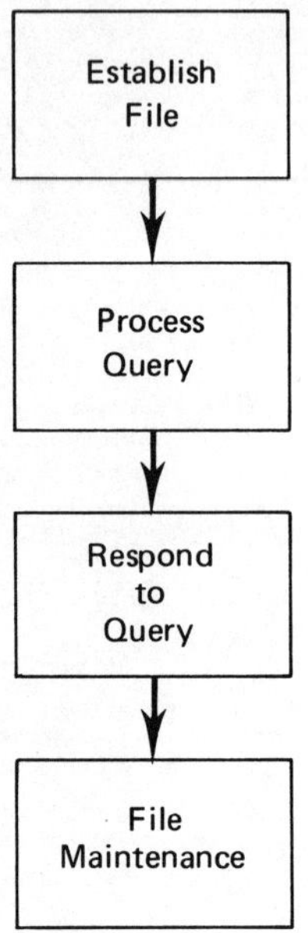


**Figure 3.9.** Types of information in a data base for a manufacturing plant

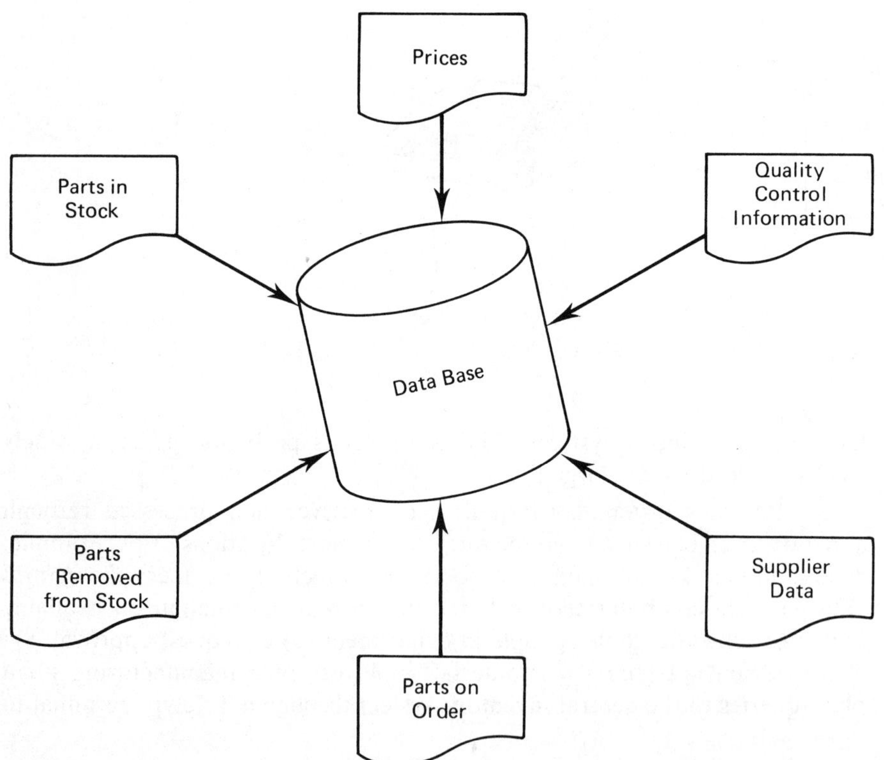

**Figure 3.10.** Data base storage media

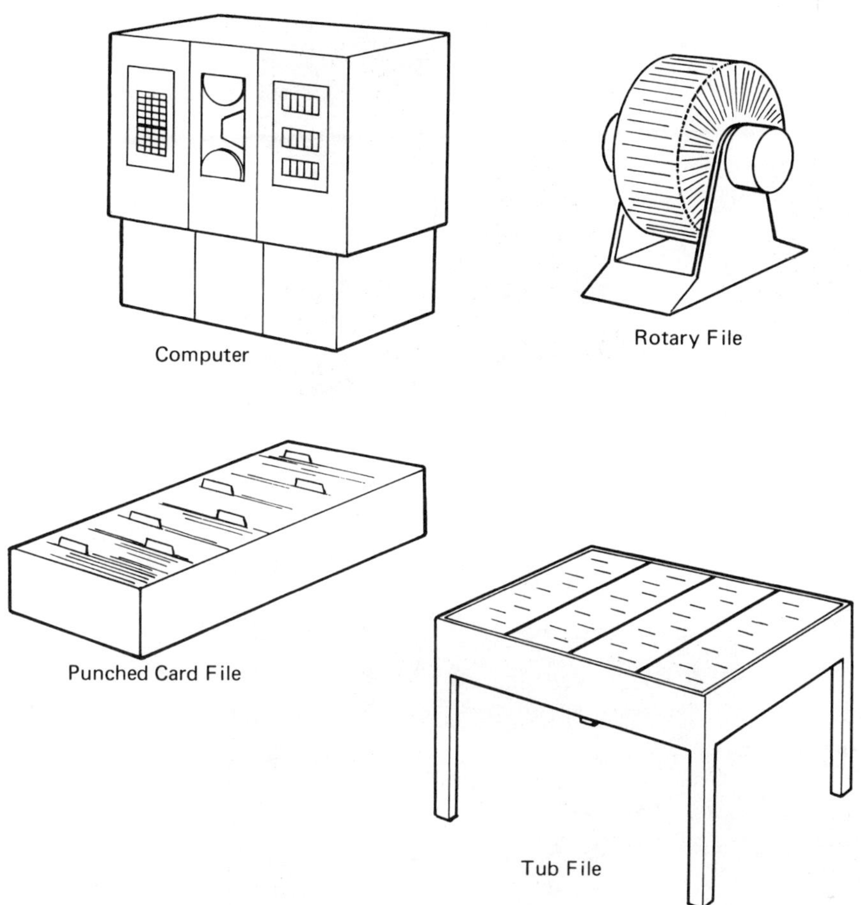


learn the number of units of a specific part in stock. Figure 3.11 shows a device used to query a data base.

3. RESPOND TO QUERY. A response or reply is made to an inquiry, giving the requested information. The data in the master file is searched, and the required item, data, or account is located. The relevant data is then communicated to the user. The type of media used for the response varies with the system. The data may be shown on a television-like tube called a video display unit (Figure 3.12) or typed out on the printer of a terminal. It may be delivered verbally by telephone or written by mail. Responding to a query might also include carrying out the actions requested, such as picking items from stock or mailing tickets.

4. FILE MAINTENANCE. As changes occur in the status of items in the data base, the file is adjusted to keep it up to date. When customers add to their charge accounts or write checks, these transactions are posted to their

**Figure 3.11.** Query terminal

Courtesy of The Singer Company.

**Figure 3.12.** Video display terminal

Courtesy of Hewlett-Packard.

account balances stored in the data base. As items are sold or taken from stock, the count in the master inventory is modified. As new units (parts, reservations, sizes) are made available, this information is added to the master file.

Maintenance is performed in different ways, depending on the system used. In a computerized system, a computer program would handle file maintenance. In a card or tub file system, corrected records would be prepared and inserted manually or by unit record machines.

### Word Processing System

This major business system is also called the text processing system. It is designed to manipulate or edit words and textual matter to generate reports, manuals, documents, letters, correspondence, and the like. It differs from ordinary data processing systems in that it processes strings of text rather than numerical data.

This type of system is used by firms that deal with a large amount of written material. Insurance companies may have word processing systems for preparation of reports and documents related to clients. Advertising firms may use the system to produce large numbers of brochures or letters, and factories may prepare training manuals with the help of word processing systems.

This system consists basically of a cycle in which textual matter (words, sentences, and paragraphs) are captured (recorded), transcribed, edited, and printed out, until a satisfactory finished document has been produced. (See Figure 3.13.) The steps in the cycle involve automated and manual processing. Dictation is usually done verbally and recorded manually in shorthand or on a dictation machine. The transcription is done on a typewriter with facilities for storing keystrokes or on a computer terminal. Editing, or revising, involves making corrections, changes, additions, or improvements in the text. The changes may be indicated manually on the draft of the document and then keyboarded. Printing out the drafts and finished documents is done by typewriter or other printing device.

**Figure 3.13.** Word processing system

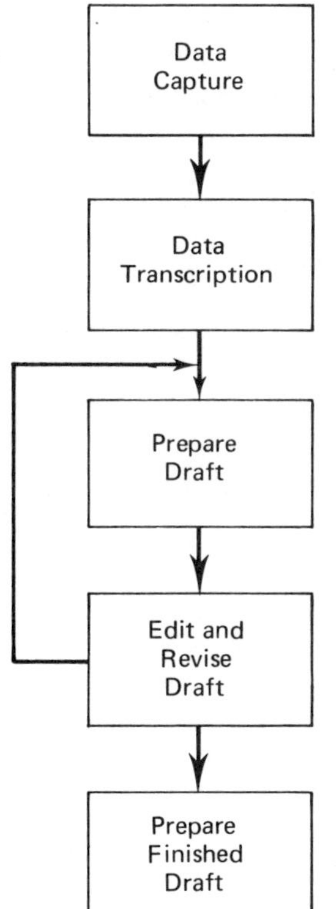


**Figure 3.14.** Dictation machine

Word processing systems involve several steps:

1. INITIAL DATA CAPTURE. The first step involves the capture, collection, or recording of textual information for further processing. The oldest method, and one still widely used today, is by verbal dictation and manual recording by shorthand or longhand. A more modern method involves the use of a dictation machine, shown in Figure 3.14. These devices record the spoken word onto magnetic belts. The belts may later be replayed at a slower speed, and stopped and started, to enable an operator to keyboard the recorded information.

2. TRANSCRIBE AND KEYBOARD TEXT. This phase involves converting the recorded verbal or handwritten text to machine readable form. In this step, an operator transcribes the shorthand notes or listens to the information recorded on the magnetic belts, and keyboards the words on a typewriter or similar device capable of saving the keystrokes. Typewriters designed for this task have some form of storage system, usually magnetic tape, magnetic cards, or paper tape, as shown in Figure 3.15. In computer systems, the keystrokes are saved on magnetic disk or tape devices. (See Figure 3.16.)

3. PREPARE CHECKING DRAFT. After the text has been keyboarded and stored, a checking draft is printed out for review. The storage media with the keyboarded data is repositioned back to the beginning, and the machine prints out the recorded keystrokes. One or more copies may be prepared during this step without requiring any further keyboarding.

4. PERFORM EDITING AND REVISIONS. At this time, the draft is reviewed and checked for accuracy and content. Changes and alterations are indicated by marking them directly on the draft.

**Figure 3.15.** Magnetic tape selectric typewriter

Courtesy of IBM.

**Figure 3.16.** Computer storage media

Courtesy of IBM.

Different methods are used to enter these changes into the stored text, depending upon the system. In a computer-based system, the changes may be keyboarded directly into memory from a terminal keyboard. The computer will insert these changes into their proper place in the stream of text matter. In a magnetic tape or paper tape system, the changes are keyboarded on a separate tape and merged with the original to produce a revised tape. Other systems enable the operator to merge corrections with the original tape, making an edited copy.

5. PREPARE FINAL DRAFT. After all additions, revisions, and corrections have been entered into the system, a finished draft is printed out. It contains all of the changes integrated into their proper place in the text stream. Sometimes, if many revisions and alterations are involved, the operator will, at this point, repeat steps three and four of the word processing system to prepare intermediate drafts.

The finished draft is frequently reformatted for improved appearance. Different line widths, spacing, or type styles may be used to improve readability. The finished document may be used as camera-ready copy for reproduction by the offset process or it may be duplicated on an office copying machine.

Word processing systems have several advantages over preparation of text matter by conventional means. They include:

1. STORAGE CAPABILITY. The capability of the hardware to store keystrokes has several important benefits. Once data is captured and entered into the system, it can be replayed to produce as many copies of the original document as necessary.

2. EASE OF ERROR CORRECTION. Most pieces of hardware allow the operator to correct typing errors by backspacing and striking over. The

corrections are automatically entered by the machine, and the draft is printed out error free.

3. Power typing. The ease of error correction allows the operator to type at maximum speed, increasing output. Operator fatigue is decreased since the operator does not have to stop to make page corrections.

4. Revisions and corrections. The system's ability to merge revisions and additions into their proper places eliminates the need for most retyping. Only revisions and corrections need to be keyboarded.

5. Formatting changes. The system allows the operator to play out drafts in different formats. Thus, different line spacing, type styles, or line widths may be used to improve page appearance or readability.

A word processing system should be designed to meet the specific needs of the particular organization. Some firms require a system that can prepare various types of reports and documents from dictation recorded in-house or phoned in from the field. Others design their systems to accommodate one person and a secretary using the same dictation/transcription device.

These variations are, in part, made possible by the type of automated equipment used in the system. Following are brief descriptions of some of the major hardware devices in use in word processing systems.

1. IBM combination unit, model 171. The desk top dictation/transcription device shown in Figure 3.17 records up to 20 minutes of dictation on a magnetic belt. Features allow rerecording to correct errors, remote control microphones, and reusable belts. The belts are replayed on the same machine for transcription.

**Figure 3.17.** Combination unit, model 171

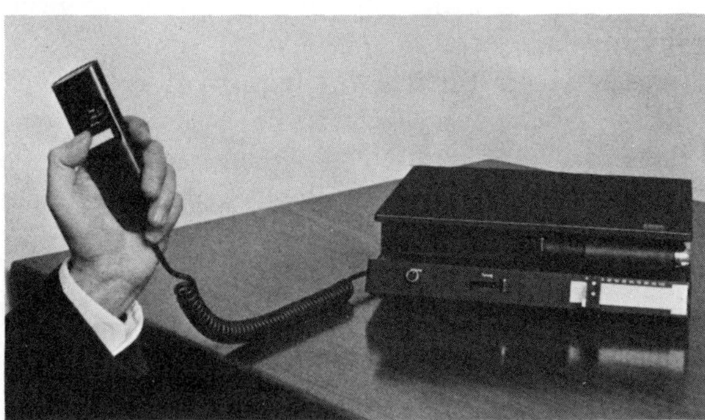

Courtesy of IBM.

2. IBM dictating unit, model 274. The small hand held dictating unit shown in Figure 3.18 is battery operated. It allows dictation to be recorded in or out of the office. Belts are removed and transcribed on a separate machine, such as the model described above.

**Figure 3.18.** Dictating unit, model 274

Courtesy of IBM.

3. IBM MESSAGE RECORDER, MODEL NO. 278. The unit shown in Figure 3.19 receives dictation from a phone line and records it on a magnetic belt. In a word processing system this unit is located in an office or data center and connected to a telephone. It can automatically answer the phone and switch on the message recorder. A user dials the telephone number of the message recorder from another office or out in the field, and is connected on-line to the device. He or she may then dictate up to 28 minutes on a belt. From a remote location, the user may stop, replay, or rerecord the dictation. Periodically, belts are removed and transcribed on a separate machine, such as the Model 271.

4. IBM MAGNETIC TAPE SELECTRIC TYPEWRITER (MT/ST). This machine (Figure 3.15) is a selectric typewriter with facilities for storing text on magnetic tape. The operator types in the original text, and the machine records the keystrokes on a magnetic tape cartridge. Mistakes in typing are corrected by backspacing to the point of error and rekeyboarding the corrected strokes. A checking draft is printed out by pressing the appropriate controls on the machine.

Changes and alterations are marked directly on the typed draft. The original cartridge is replaced in the machine and the text printed out to the point where an addition or revision is to be made. The operator keys in the change and the machine moves ahead and prints out the text to the next point of error. (See Figure 3.20.) On some models a new corrected tape is generated during this step. When all changes have been entered, a finished document is printed out from the new tape. On other models, the operator

**Figure 3.19.** Message recorder, model 278

Courtesy of IBM.

**Figure 3.20.** Word processing procedure on the MT/ST

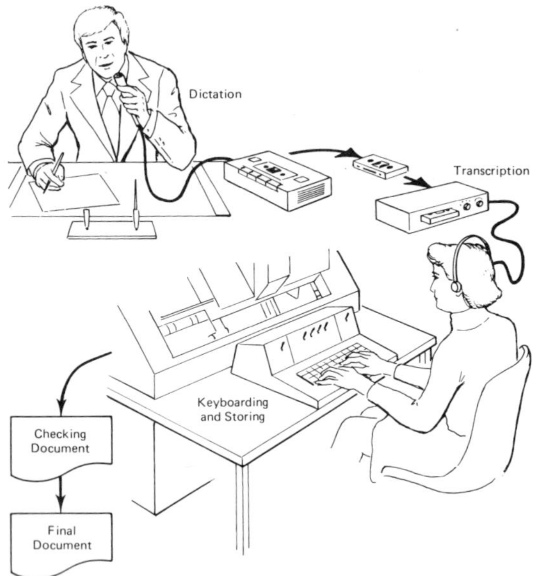


keys in the changes and additions in the appropriate places as the finished draft is produced.

The machine gives the operator the advantage of selecting different line widths, spacing, and type styles when preparing the finished document.

5. IBM MAGNETIC CARD SELECTRIC TYPEWRITER (MC/ST). This system, shown in Figure 3.21, is similar to the magnetic tape selectric typewriter, except that keystrokes are recorded on a magnetic card. The card measures the same as a standard 80-column punched card, 3¼″ × 7⅜″. Each card holds approximately 5000 characters, enough for an entire letter or memo. Cards may be filed, sorted, mailed, and indexed. The final draft may be printed out in a line size, type style, and spacing different from that of the original recording.

**Figure 3.21.** Magnetic card selectric typewriter

Courtesy of IBM.

6. PAPER TAPE SYSTEM. Paper tape is a common media used to store keystrokes in a word processing system. Paper tape facilities are found on terminals and some typewriter machines. (See Figure 3.22.) A coded com-

**Figure 3.22.** Paper tape

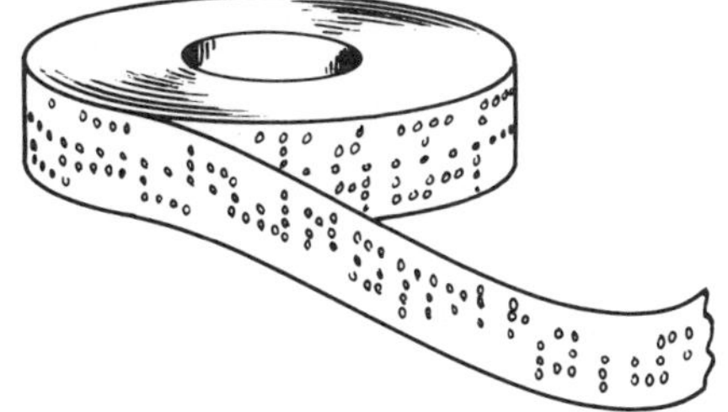

bination of holes is punched into the paper tape as the operator keyboards the characters. The punched paper tape is placed in the tape reader, and a checking draft is printed out. Alterations and additions are made by printing out the characters on the paper tape to the point of error, entering the changes, and moving ahead to the next point. A second corrected tape is often punched at the same time.

7. MINICOMPUTER SYSTEM. Minicomputers such as the one shown in Figure 3.23, are often used in word processing. These systems have magnetic disk or tape facilities for storing text. Errors made during typing are corrected by such means as reentering lines or backspacing and retyping. The computer will print out a checking draft incorporating these corrections. Additional changes and alterations in the text are keyboarded into the storage system, and the computer automatically places them in their proper places in the text. The finished document may be printed out following the same or a different format.

8. LARGE SCALE COMPUTER SYSTEM. Large computer systems, capable of holding millions of characters in storage, are also used in word processing. Up to several hundred terminals may access the system at one time. This enables many operators to generate text copy, reports, letters, or memos simultaneously.

In this type of system the operator first opens a file which directs the computer to reserve storage space. Then the operator keyboards in the

**Figure 3.23.** DEC PDP 8 minicomputer

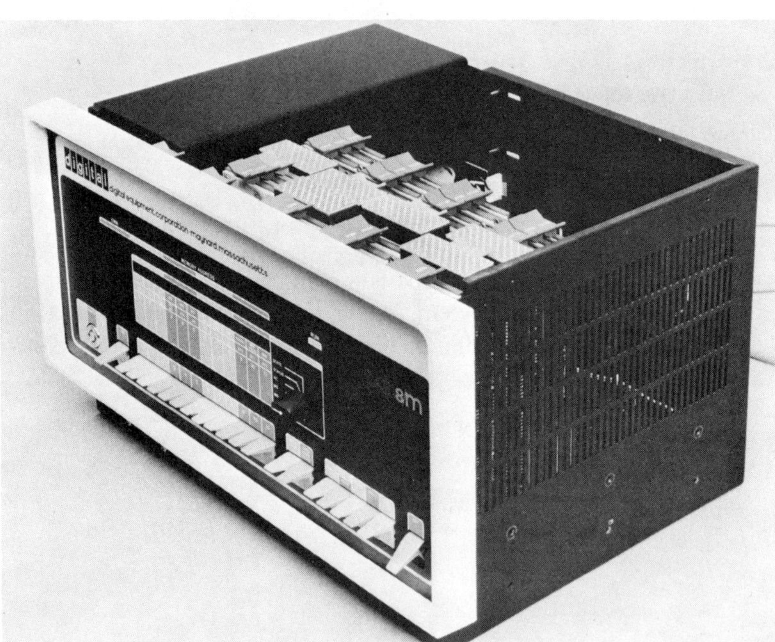


Courtesy of Digital Equipment Corp.

original text and instructs the computer to print out a checking copy. Revisions and alterations are entered and inserted into the correct positions by the computer. When the final draft is approved, a finished document is printed out.

9. TIME SHARE WORD PROCESSING SYSTEM. Word processing services are also available from time share firms on a per hour basis. Remote terminals, tied to the computer via ordinary telephone lines, provide the communications link. A terminal is installed in the user's office. He or she dials the number of the computer, logs on, and opens a file to save the keystrokes entered from his or her terminal. Drafts are printed out on the print unit of the remote terminal. Changes and alterations are entered from the terminal and inserted into the proper points in the text by the computer. The finished, error-free document is printed out on the remote terminal.

### Management Information System (MIS)

The primary purpose of this system is to provide management with a diverse selection of current information regarding the status of the business for use in making decisions. The management information system is similar to the inquiry system described earlier, but is more comprehensive. The inquiry system involves maintenance of a master inventory file. The MIS involves making information on the different aspects of the business available for decision making and study.

The system is built around a data base that contains information on customers, sales, inventory, finances, personnel, distribution, orders, and such. This file is structured to allow selected information to be accessed. The information is restructured in various ways to provide a variety of reports for management. The reports may be generated routinely at regular intervals to show daily fluctuations and activities. They may also be generated at the specific request of a user and contain only selected information. Figure 3.24 shows the organization of information in a data base.

Developing the data base for an MIS usually follows the steps described below:

1. ESTABLISH THE DATA BASE. In this initial phase, a study is made of the elements to be included in the data base. Types of information and records to be included are defined, and the manner in which the system will be accessed and maintained is planned. The type of output reports are described and designed.

The layout of the records is one of the important elements to be considered when planning the data base. The records must be consistently structured to allow maximum utility and access to the recorded data. The columns on the records are often assigned so that data related to the same category is recorded in the same columns on each record, regardless of what kind of record it is.

**Figure 3.24.** Data base construction

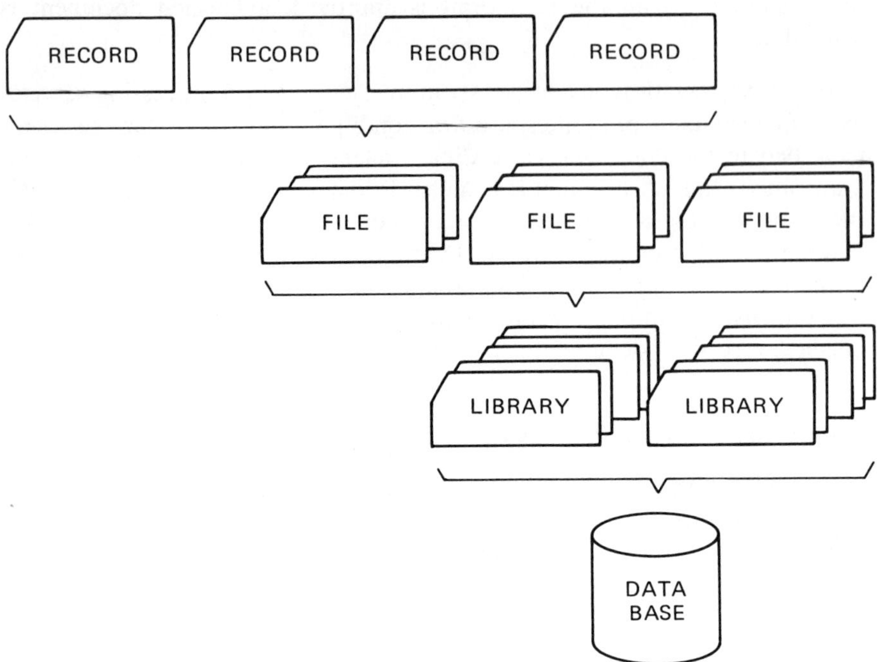


For example, the department number may be recorded in column 26 on all records whether they be from the personnel, inventory, or purchasing departments, or related to expense accounts or sales invoices. At any time, all information relating to a specific department can be accessed by sorting column 26.

2. GENERATE OUTPUT. Periodically, reports are generated for management utilizing the data in the data base. Some reports may compare corresponding data in different departments or time periods. Others may show sales versus costs or changes in the status of a specific variable. These reports may be outputted in several ways—some in permanent form on a line printer or terminal print unit; others displayed temporarily on a video display terminal. The data base should also be able to generate, on request, other reports showing selected data and comparisons.

3. MAINTENANCE OF THE DATA BASE. As new information is collected on business activities, the status of the data base must be modified to reflect these changes. Outdated information is replaced; new data, facts, and figures are entered. Changes in the data base are made either from a central location or from a remote terminal.

## TELEPROCESSING SYSTEMS

Many of the systems so far described utilize communications links to transmit data from one location to another. This arrangement is called a teleprocessing system. Such systems are composed of one or more computers and terminals situated in different locations, tied together by a communications link. Telephone lines, microwave circuits, and private wire circuits are among the most common communication links used. Information, reports, and other data are sent from one station in the system to another via the communication link.

A teleprocessing system has four basic functions:

1. DATA MOVEMENT. In this arrangement, large quantities of data are moved from one station in the system to another. This could be from one terminal to another, from one computer to another, or from a computer to a terminal, depending on the design of the system.

2. LINK PROCESSING AND DATA INPUT ELEMENTS. Teleprocessing systems can be used to tie the stations where data is input to the locations where it is to be processed. In this arrangement, teleprocessing provides the link between the locations where the source data is generated or prepared for computer input (sales floor, bank teller terminal, stock room) and the computer at the data center where it will be processed and manipulated.

3. MULTIPROCESSING. Teleprocessing systems are used to tie together two central processing units (CPUs) so that their facilities can be shared. This might be done to make data stored in one data center available to another; to provide backup services; or to relieve congestion on one machine during peak rush hours.

Using teleprocessing systems, businesses can take advantage of the time differences across the country. The New York branch of a firm can access computers in California before the working day begins in the West. Conversely, the California branches can use the CPUs on the East Coast after their working day has ended.

4. MESSAGE SWITCHING. In this arrangement, a computer functions as a central switchboard. It receives data from a terminal or another computer on the system, and relays it to the appropriate station. A business with branches in different parts of a city or the country may use a teleprocessing system to convey communications.

The teleprocessing system itself is the focal point of other businesses. Organizations such as news bureaus, in which the movement of large quantities of data is the main activity (rather than supportive of the main activity), are an example of this type of usage. All business functions and data flow procedures and activities are centered around the teleprocessing system.

The following are examples of some typical teleprocessing systems:

1. SIMPLE TELEPROCESSING SYSTEM (Figure 3.25). A central office houses the computer with one remote terminal or inquiry station at another location. The terminal and the computer are tied together through a communications link.

**Figure 3.25.** Simple teleprocessing system

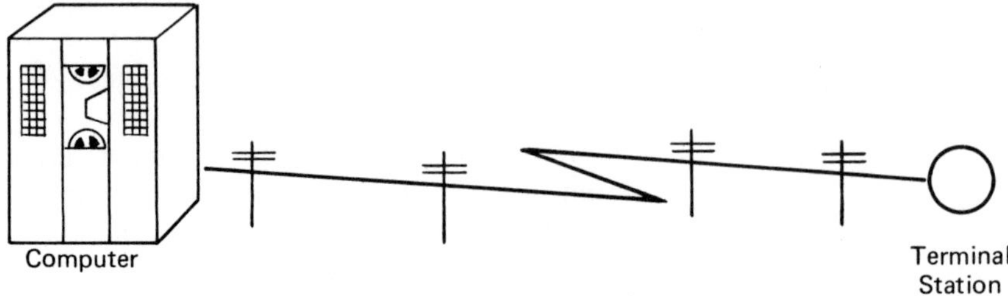


A small tax accounting firm might use this type of system. Because of its limited size and resources, the company may prefer to buy processing time from a time sharing firm, rather than to purchase its own computer. A terminal, located in the accounting firm's office, is connected to the time sharing computer via ordinary telephone lines. Each day information is transmitted over the communications line to the computer, where it is processed, and the results returned in the form of a report on the remote terminal.

2. MORE COMPLEX TELEPROCESSING SYSTEM (Figure 3.26). This system has several terminals tied to a central computer, enabling several users to share the resources of a single computer. A bank with several branch offices located in a city might use this type of teleprocessing system. A computer is located in the main office, with several teller terminals in each branch. Customers' transactions are keyboarded promptly by the tellers as the customers wait at the windows. The information is relayed to the central computer to update the account. A customer's passbook is posted automatically on the terminal or manually by the teller.

Users can have access to the computer throughout the day, at the moment a transaction occurs. This is an example of a system that can perform online, real time teleprocessing.

3. COMPLEX TELEPROCESSING SYSTEM (Figure 3.27). This illustrates a large, complex teleprocessing system. It is composed of a network of many remote terminals and several computers. Such a system might be used by a national credit and collection agency which provides up-to-the-minute credit information on commercial accounts. Each major office or branch has its own computer with several connecting terminals in the local offices. The branch office computers are tied together by the teleprocessing system. This allows a pooling of resources and an exchange of information

**Figure 3.26.** More complex teleprocessing system

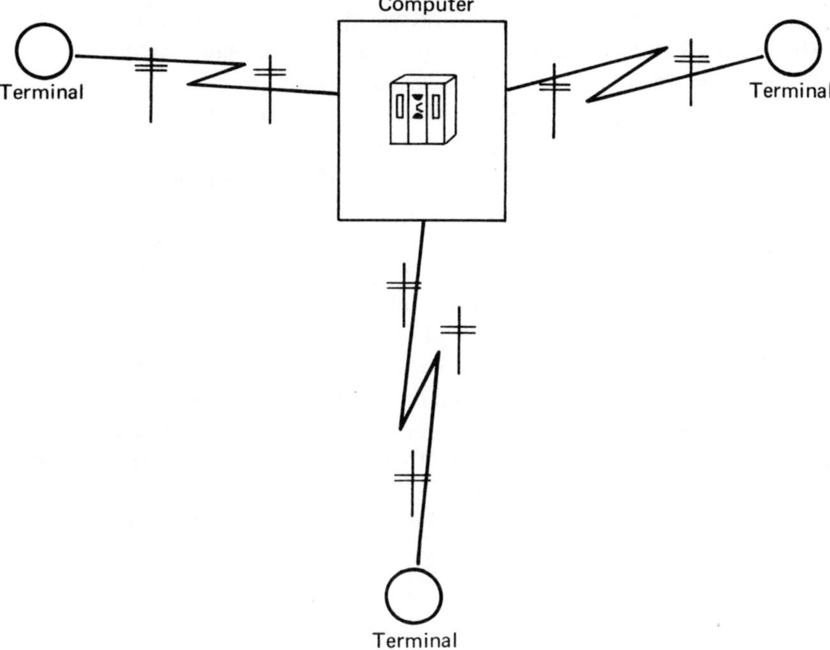


**Figure 3.27.** Complex teleprocessing system

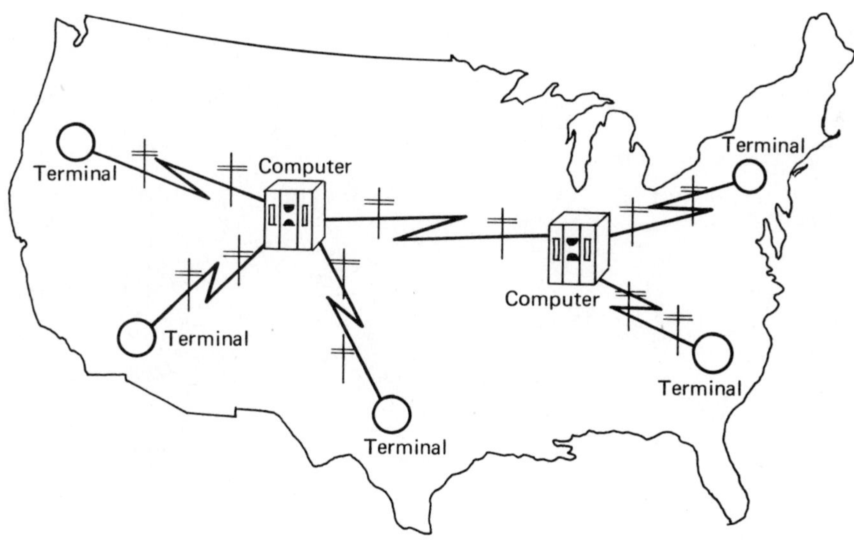


on a national basis. This type of system may also be used by large insurance companies, retailers, or manufacturers.

## EXERCISES

1. Describe the purpose of the order processing system.
2. Summarize the steps followed in a typical order processing system.
3. Describe the function of an inquiry system.
4. Summarize the major steps involved in the inquiry system.
5. Describe the function of the word processing system.
6. Summarize the major steps involved in the word processing system.
7. How does the word processing system differ from a data processing system?
8. Select one of the major pieces of word processing system equipment described in the chapter and summarize its features and functions.
9. Describe the functions of a management information system.
10. Summarize the major steps involved in a management information system.
11. Describe the functions of the teleprocessing system.
12. Summarize the major steps involved in a teleprocessing system.
13. Select a business firm to study. Interview the proprietor or an employee. Determine whether the system falls under one of the major types discussed in this chapter. Describe its major features.
14. Visit a bank that uses a computer facility, and describe the type of system they use.
15. Visit a small manufacturing firm that uses an order system, and describe its major features.

# Chapter 4

# Data Processing Methods

One of the most important decisions a systems analyst makes when designing a business system is deciding when to use a ten-cent pencil and when to use a million-dollar computer. The choice is not as simple as it seems. Many factors are involved—volume of data to be processed, demand for accuracy, system reliability, availability of human resources, cost of equipment and personnel, and others.

It is the responsibility of the systems analyst to be familiar with the elements of modern data processing methods; to know the advantages and limitations of each; and to decide which will handle a particular problem in the most efficient and effective way.

This chapter describes the three major methods of data processing in use today—manual, unit record, and computer. Most modern business systems use one or more of these methods, alone or in combination.

## THE DATA CYCLE

The three major methods of data processing vary greatly in many aspects—technology, expense, speed, capacity—but they all move data through the same three basic steps during processing:

INPUT ⟶ PROCESSING ⟶ OUTPUT

This sequence of operations is often called the data cycle.

The first step consists of capturing, or entering, data into the system. Information recorded on source documents is converted, if necessary, into a form suitable for processing by the method used, and entered into the system. Source documents refer to original information recorded, often longhand, on sale slips, deposits slips, time cards, or order forms. This data may be prepared for input by keypunching, keyboarding, or copying longhand into special ledgers or forms. In some instances, the source documents themselves are suitable as input data.

The second step involves restructuring the data to make it more useful to the enterprise. Most common data processing activities involve one or more of the following operations:

sorting, sequencing, and alphabetizing
classifying
storing
collating and merging
searching and selecting
manipulating arithmetically
converting into a form suitable for further processing by the same or another method

Each method of data processing is capable of performing these processing operations. The time required and the degree of accuracy achieved will vary, of course, with the method used and the volume of data being processed.

The last stage in the data cycle consists of outputting, or delivering the results of processing. In this step, the processed data is converted into a form useful to people, such as printed reports or documents. Examples include paychecks, statements, invoices, rosters, and lists. Converting processed data into a form suitable for storage or later input to the same or a different system is another major type of output. Examples include updated accounts receivable files stored on magnetic tape or punched into cards; sales data restructured for use in preparing the payroll or for retrieval in a management information system.

## MANUAL DATA PROCESSING

Manual data processing involves manipulating data at the speed of the human hand and eye. Pencil and paper are the principle aids. Mechanical or electrical devices, such as the typewriter, abacus, slide rule, adding machine, mechanical or pocket electronic calculator, are often used in processing. (See Figure 4.1.)

Source data is prepared for input to the system by writing it in longhand, keying it on a typewriter, or manually entering it into a calculator. Processing is performed on paper, mentally, physically, or manually on a calculating device.

**Figure 4.1.** Manual method

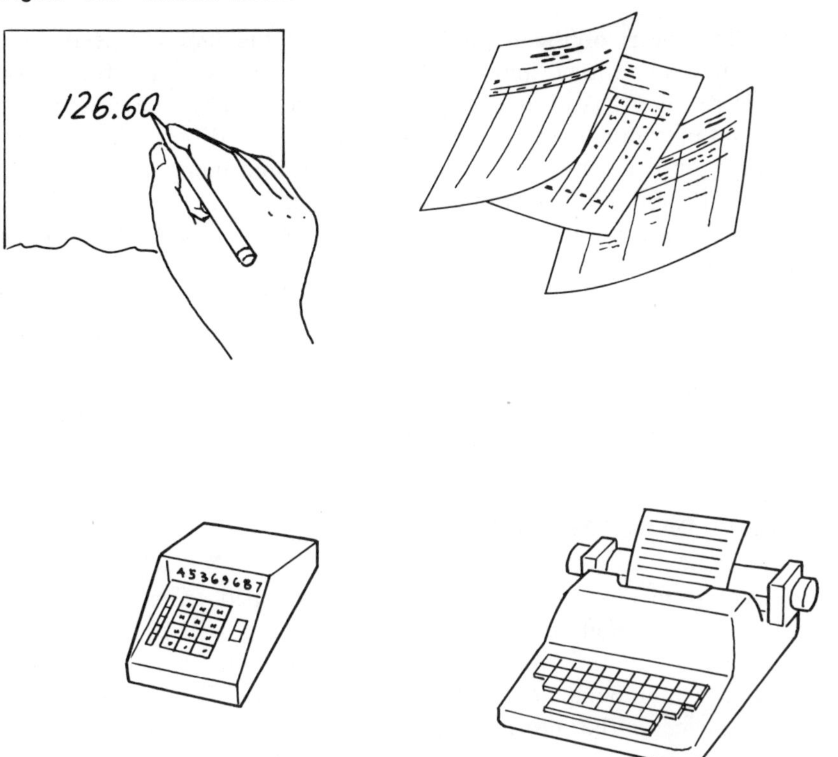


Lists are alphabetized on cards or on paper. Figures are transcribed by hand from sales slips to ledgers, and totalled on an adding machine. Specific records are pulled from a file by hand.

Output consists of handwritten or typewritten information or documents. Examples are paychecks, memos, reports, statements, invoices, and ledger sheets.

### Applications

The manual method of data processing is widely used in small businesses, when small amounts of data are involved or when there is little time pressure on the data cycle. It is also often used when a particular sequence of data processing procedures will be performed one time only.

### Advantages

The advantages of the manual method include simplicity, low cost of equipment, and the ease of training personnel and making modifications in the procedures. Another favorable factor is that the required aids and equipment are readily available.

### Limitations

The major limitations of the manual method are the limited speed and accuracy of the human hand and eye. People can process only a few hundred records per hour. And as speed increases, so usually does the error rate. A human operator makes about one mistake per 100 calculations—a relatively high rate. (A modern high-speed electronic computer averages one error per 10,000,000 operations.)

Lack of standardization is another limitation on the manual method. People are unique, and everyone approaches a problem a little differently. This makes it difficult for someone else to check the work of a human operator or to trouble-shoot a manual system. Another related limitation is that people do not always perform a procedure exactly the same way each time they do it. This leads to variations in output, complicates training employees, and adds to the difficulty in locating any errors.

## UNIT RECORD DATA PROCESSING

The unit record method of data processing developed from a machine designed by Herman Hollerith in 1889. Hollerith, employed by the United States Census Bureau to process the 1890 census, built a machine that could read and manipulate data punched into cards. The cards moved through the device as the information punched into them was read, processed, and tallied.

The modern unit record system still centers on the punched card and uses a variety of machines to perform many different processing procedures on the encoded data. Since most of these machines are automated, the unit record system is also known as Electrical Accounting Machine processing, EAM processing, or, sometimes, punched card accounting. Small computer systems are being utilized more and more in modern unit record processing and are supplanting the older automated machines.

Each punched card is a record containing data related to one transaction. The data is recorded in the card in the form of punched holes. Each letter, number, and special character has its own combination of holes.

Figure 4.2 shows the 80-column card, which is the most widely used punched card. Figure 4.3 illustrates the standard 12-punch position code used to punch characters into this card. (It is based on the code system originally developed by Hollerith.) Data is input, or recorded, on the card by a device such as the cardpunch shown in Figure 4.4. As a key is depressed, the appropriate combination of holes is punched into the card.

Figure 4.5 illustrates the newer 96-column punched card. The code for this card, using six punch positions, is shown in Figure 4.6. The data

**Figure 4.2.** 80-column punched card

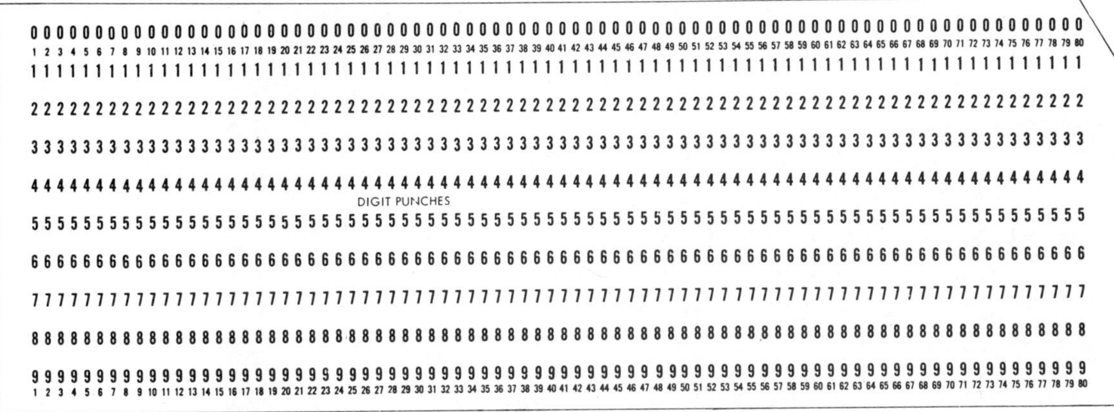


Courtesy of IBM.

recorder shown in Figure 4.7 is used to record data in the 96-column card.

The English translation of the punched data can be printed at the top of the card by the punch machines to facilitate human interpretation.

Processing in the unit record system is done with a group of machines that are capable of reading the data punched into cards and manipulating it in various ways. The punched cards are manually loaded into a unit record machine and moved mechanically through the device. Reading brushes or lights sense the hole combinations punched in the cards. Unit record machines do not read the information written or printed on the face of the cards.

Many unit record machines are programmed, or controlled, by wiring

**Figure 4.3.** Standard punch code for 80-column card

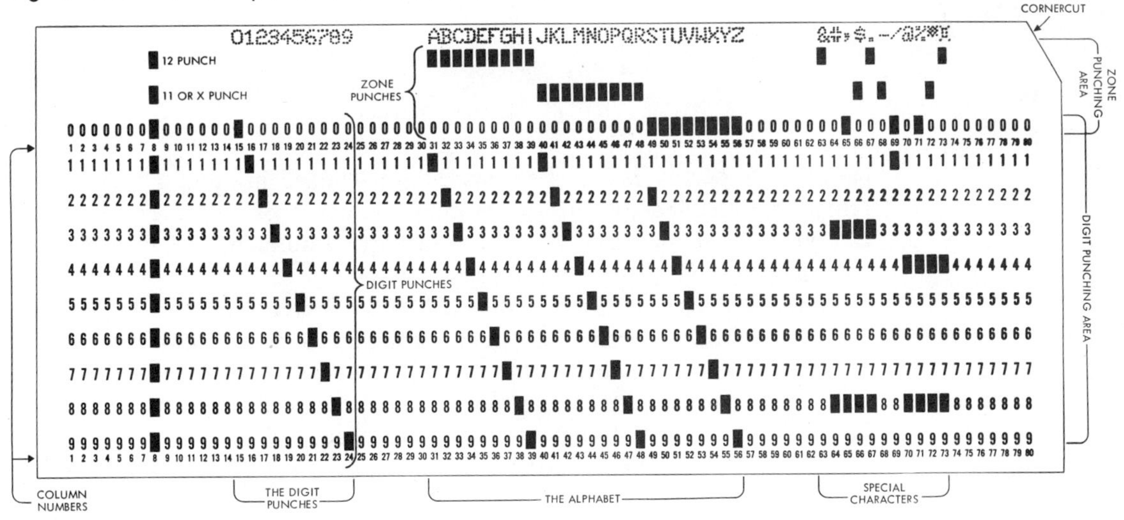


Courtesy of IBM.

**Figure 4.4.** 029 card punch

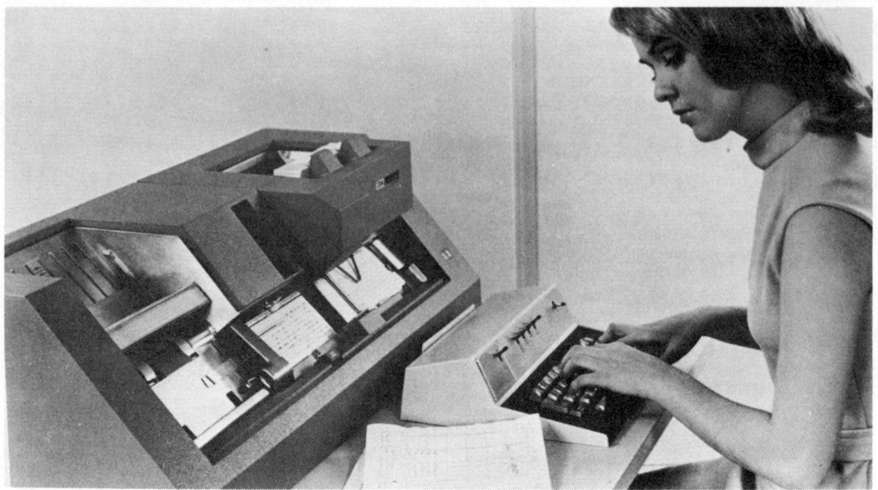

Courtesy of IBM.

**Figure 4.5.** 96-column punched card

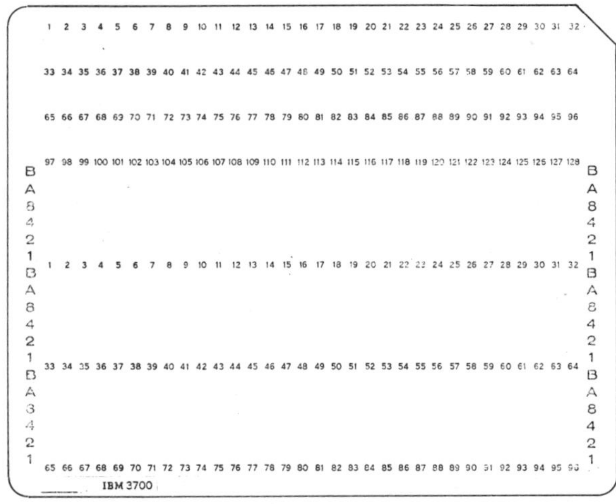


boards. Jumper wires on these boards are plugged in certain positions to direct the machine to perform specific sequences or steps. The machine can be re-programmed or directed to perform a different sequence of steps by changing the positions of the jumper wires.

A sorting machine, such as the IBM 084 Sorter shown in Figure 4.8, will sort cards according to the information punched into any of the columns. This device feeds cards from a hopper, reads the punched data, and drops the cards into one of several receiving pockets. Thus, cards may be sorted by numerical sequence or alphabetized.

The IBM 085 Collating Machine (Figure 4.9) is designed to match files

**Figure 4.6.** Standard punch code for 96-column card

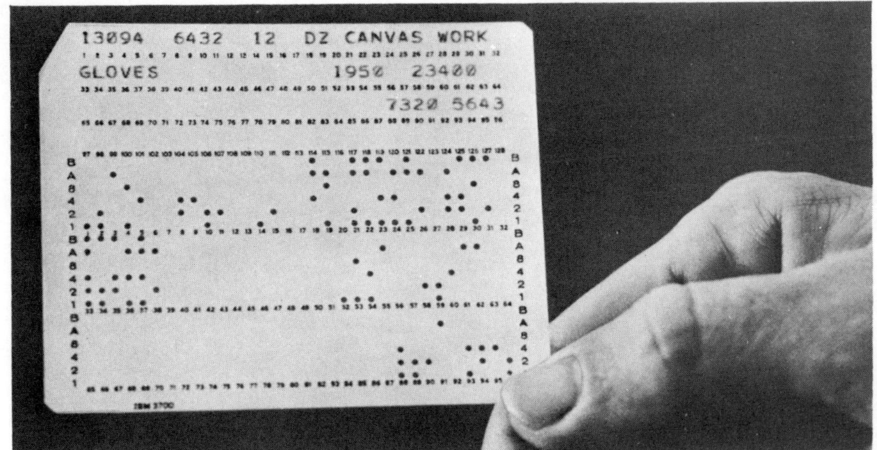


Courtesy of IBM.

**Figure 4.7.** 5496 data recorder

Courtesy of IBM.

**Figure 4.8.** 084 sorter

Courtesy of IBM.

**Figure 4.9.** 085 collating machine

Courtesy of IBM.

or merge several files into one. The machine can match related cards from two files, pull aside cards without corresponding matches, check cards for sequence, or select cards from a file.

The IBM 519 Document Originating Machine (Figure 4.10) is designed to handle a variety of unit record operations. It can duplicate the information punched in one file of cards into another set of cards. In 80-80 reproducing, all 80 columns of information from one card are copied into the corresponding 80 columns of another card. In offset reproducing, the machine copies data from one card into different columns in another card. It can generate a master card and several detail cards in gang punching.

The IBM 519 can process mark sense cards, which are a method of data entry. It can read coded data pencilled on a card and convert it into punched holes that can be read by other machines. The IBM 519 will also do end printing. In this operation, punched data is interpreted and printed in large letters at the end of the card.

The IBM 548 Interpreting Machine is designed to translate the holes punched into a card and print the English equivalent on the card's face. The print unit is capable of interpreting up to 60 of the 80 available columns on the standard card.

The IBM 609 Calculator shown in Figure 4.11 is used to process data in the unit record method. It is capable of reading data punched into cards; storing a small amount of data in its memory; adding, subtracting, multiplying, and dividing the stored data; and punching out updated information into new cards.

Data output in the unit record method is principally performed on the IBM 407 Accounting Machine (Figure 4.12), which is equipped with a

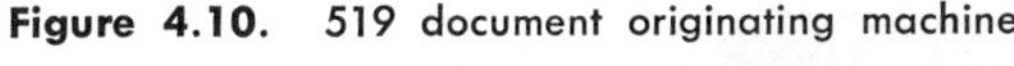

**Figure 4.10.** 519 document originating machine

Courtesy of IBM.

**Figure 4.11.** 609 calculator

Courtesy of IBM.

**Figure 4.12.** 407 accounting machine

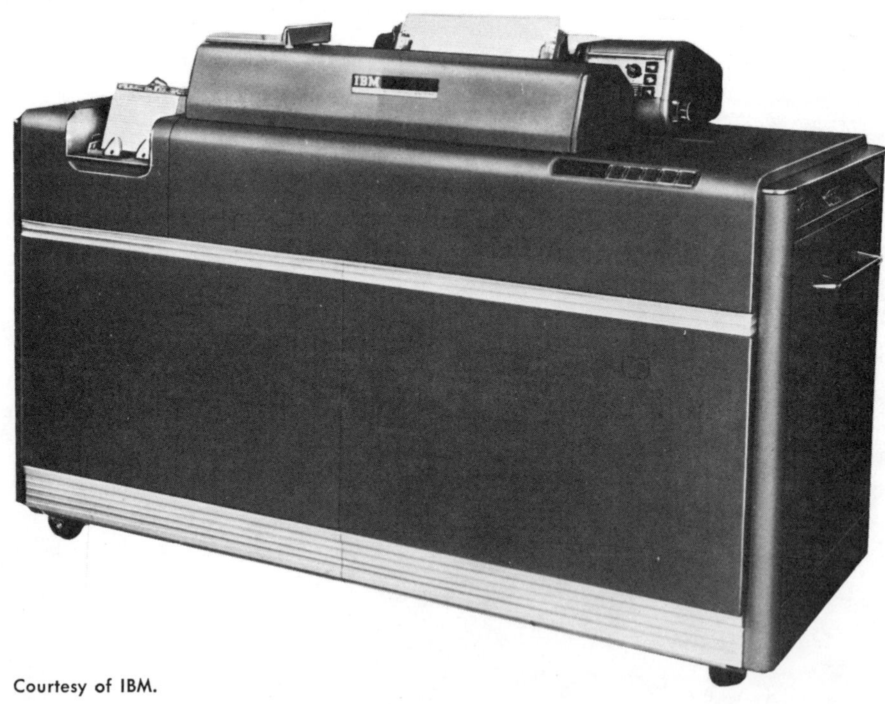


Courtesy of IBM.

card hopper and a print unit. It reads information from punched cards and prints out the data on a continuous form. It is capable of performing summary addition and subtraction on input data. The accounting machine is used to prepare such forms as statements, paychecks, invoices, or ledger cards.

The UNIVAC 90/30 system (Figure 4.13) is an example of a complete computer data processing system. It is capable of inputting, processing, and outputting data. The system includes disk storage, a card reader, processor, and printer. Magnetic tape units peripheral devices give it additional processing and storing capabilities.

The processor can perform addition, subtraction, algebraic and alphabetic comparisons, data transfers, and branching. The UNIVAC 90/30 is controlled by a connection panel. Data can be output in the form of printed documents or reports, and/or punched into cards.

### Applications

Punched card data processing systems were widely used by large business firms before the advent of the electronic computer. Since many unit record machines are readily available and their cost is relatively low, this method is still an important means of modern data processing.

Punched cards are a convenient media. They can be easily filed, handled, mailed, written on, sorted, and stored in limited quantities. They are also one of the major means of inputting data into a computer system.

Punched card processing is suitable for handling small to medium

**Figure 4.13.** UNIVAC 90/30 system

Courtesy of Sperry Univac, a division of Sperry Rand Corporation.

amounts of data. Practical applications include stock and inventory systems, payroll systems, accounting systems, and address lists.

### Advantages

The unit record method has a higher degree of accuracy than do manual methods of processing. Since machines perform most of the processing, the level of standardization is greater than in the manual method. A processing procedure will be done exactly the same way each time it is performed, regardless of who is the operator. The ease of programming unit record machines increases their flexibility by enabling one machine to perform several different procedures.

### Limitations

Relatively slow speed and the bulk storage requirements of punched cards are the major limitations of the unit record method. The machines can process from 200 to 2000 cards per minute, which is appreciably slower than computerized methods.

Punched cards are not an ideal media for storing large amounts of data. Files of such cards are bulky—a stack 2000 deep is about 20 inches high. Punched cards are easily mutilated or torn, and damaged records will not run through the machines properly. The unit record method relies on the physical and mechanical movement of cards. They must be physically carried from one machine to another, loaded, and mechanically driven one by one through the device.

## COMPUTER DATA PROCESSING

Computerized methods of data processing are an essential part of modern business systems. Their speed, level of accuracy, and flexibility of processing are unmatched by any of the other methods. These facilities are possible because computers manipulate data in the form of electronic pulses at the rate of millions of pulses per second.

Input data is converted from whatever form it is recorded in (punched cards, magnetic tape) into electronic pulses and entered into the computer. These pulses are manipulated within the computer in electronic circuits, and then converted back into one or more forms of output. (See Figure 4.14.)

A computer system is composed of several major blocks, or subsystems, which work together interactively to process data. (See Figure 4.15.)

Figure 4.14. Computer data cycle

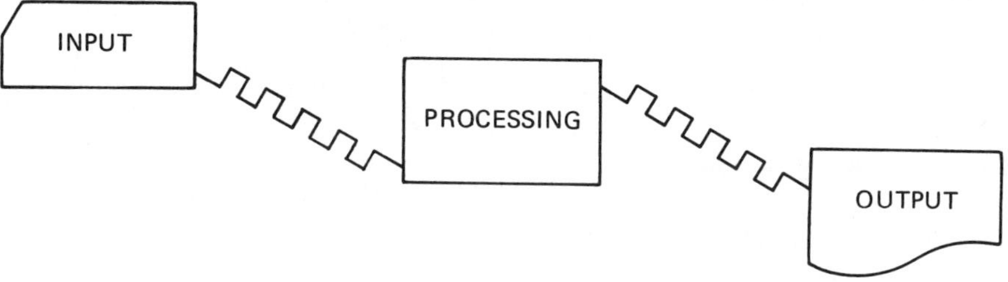


Figure 4.15. Computer system

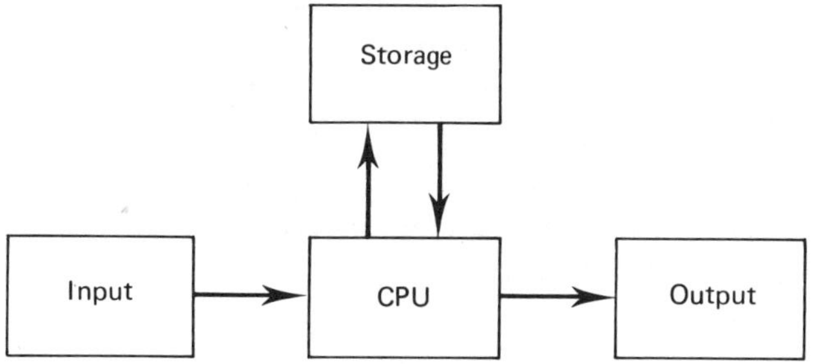


## Central Processing Unit (CPU)

The CPU shown in Figure 4.16 is composed of three sections: arithmetic and logic unit; control unit; and the primary or core storage unit. The control section monitors the overall operation and timing of the system. The arithmetic and logic unit performs mathematical calculations and numerical manipulations on the data, as well as algebraic and logical comparisons.

Figure 4.16. Central processing unit

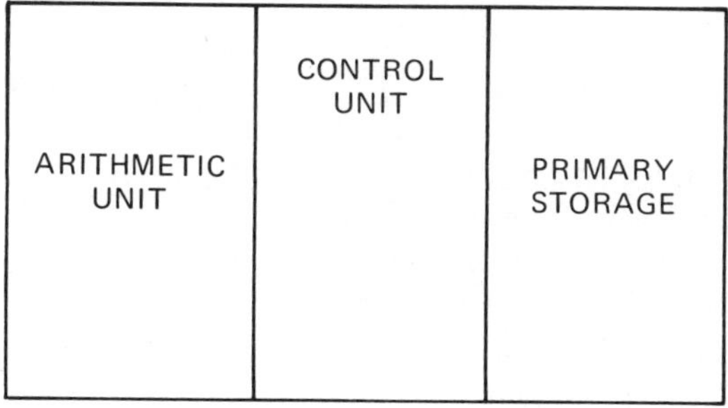


The primary storage unit is composed of thousands of tiny ferrite cores, or other magnetic storage devices, such as plated wire, thin film, or monolithic storage. Areas on these devices are magnetized to represent data. The data is held in storage until needed and can be read or copied into other parts of the computer system at the rate of millions of pulses per second. The primary storage system holds many thousands of characters or digits, ready for instant retrieval or processing.

THE STORED PROGRAM. The ability of a computer to store or remember large quantities of data gives it its power and flexibility. A series of instructions, called a program, can be written, read into the computer, and stored in the primary storage unit. The instructions direct the computer to perform a series of procedures on other data in storage. They direct it to perform mathematical operations, move or copy data, make logical decisions, branch, and input and output data. The computer will execute the instructions in the stored program, one at a time, on the appropriate data.

The program can be written and tested before the actual data is ready to be processed. It can be executed over and over on different sets of data. The stored progam is executed automatically by the computer, without direction or intervention from the programmer or machine operator.

INPUT SYSTEM. The computer's input system is designed to receive data and convert it into electronic pulses suitable for computer processing. Several common media are used for computer input:

1. Punched card. The 80- and 96-column cards are an almost universally used media for computer input. The hole combinations punched into the cards are converted into electronic pulses by the card reader and sent to the CPU for processing.

2. Paper tape. Information punched into roles of paper tape is converted into electronic pulses by the tape reader and sent to the CPU for processing.

3. Magnetic tape. Information is recorded by magnetizing areas on the surface of the tape, which is similar to ordinary domestic recording tape, to represent the binary code. These magnetized areas are sensed by the magnetic tape reader, converted into electronic pulses, and input to the computer for processing.

4. Keyboard. The keyboard, similar to a typewriter, may be on a remote terminal connected to the CPU via a communications link, or on a local terminal with a direct connection. It may also be the console typewriter attached directly to the CPU. (See Figure 4.17.)

The input system of the computer senses the keystrokes as each key is depressed, and sends a corresponding electrical signal to the CPU for processing. Keyboarding is a relatively slow means of data entry, and is generally not used except for a limited amount of data.

5. Optical character recognition devices (OCR). An OCR device (Figure 4.18), which can read handwritten numbers and special characters as well as several typewriter faces, is often used for data entry. The OCR device optically senses the shape of the letter form and converts it into pulses which are fed to the CPU for processing. This method is used to

**Figure 4.17.** 530 computer with keyboard

Courtesy of Xerox Corp.

**Figure 4.18.** 1287 optical reader

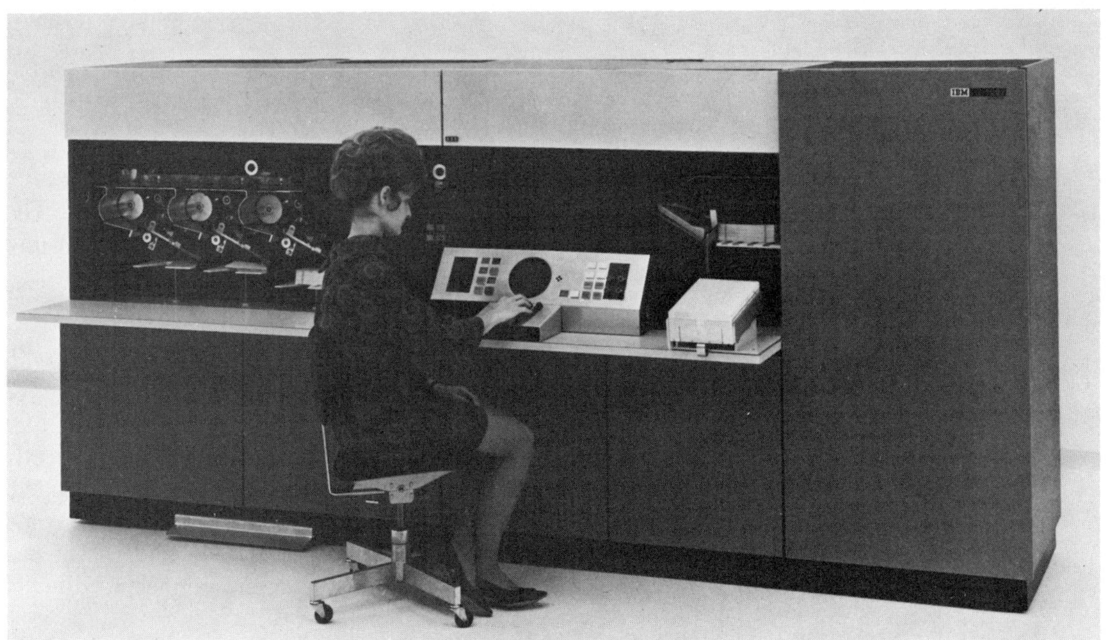


Courtesy of IBM.

read checks, sales slips, adding machine and cash register tapes, and the like. Figure 4.19 is an example of handwritten numbers that can be read by an OCR device.

**Figure 4.19.** Input for OCR

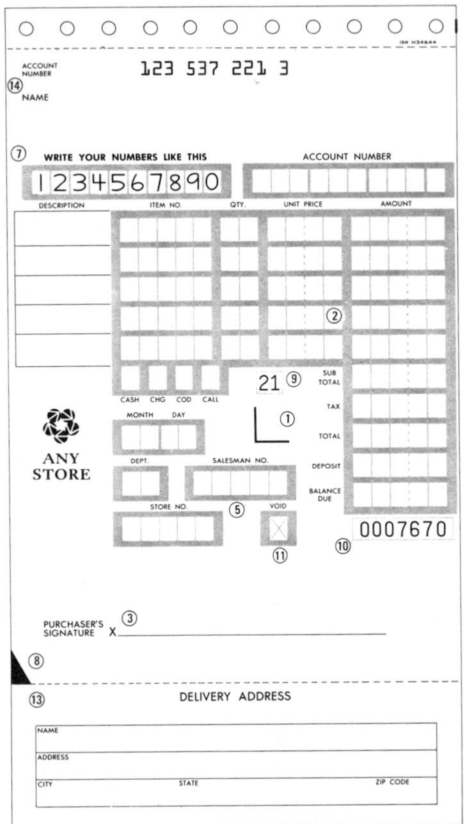
ACCOUNT NUMBER 123 537 221 3
(14) NAME
(7) WRITE YOUR NUMBERS LIKE THIS
1234567890
ACCOUNT NUMBER
DESCRIPTION
ITEM NO.
QTY.
UNIT PRICE
AMOUNT
(2)
21 (9)
SUB TOTAL
CASH CHG COD CALL
MONTH DAY
(1)
TAX
TOTAL
ANY STORE
DEPT.
SALESMAN NO.
DEPOSIT
BALANCE DUE
STORE NO.
(5)
VOID
0007670
(10)
(11)
PURCHASER'S SIGNATURE X (3)
(8)
(13) DELIVERY ADDRESS
NAME
ADDRESS
CITY STATE ZIP CODE

6. Magnetic Ink Character Reader (MICR). Data is encoded on forms, such as checks and deposit slips, with a special magnetic ink. The device (Figure 4.20) senses the magnetized characters, converts them into corresponding electronic pulses, and sends them to the CPU for processing. Figure 4.21 shows a deposit slip used in MICR data entry.

SECONDARY STORAGE SYSTEM. Most computers are equipped with large secondary storage systems that greatly expand their available storage capacity. Secondary storage systems hold large files, programs, and other data not immediately needed by the CPU. The stored data is directly accessible by the CPU and is called into active (primary) storage as needed.

Secondary storage devices include magnetic tape, magnetic drum, data cell, and magnetic disk storage systems. These devices hold millions of bits of information, recorded magnetically in binary, and are connected directly

**Figure 4.20.** 1255 magnetic ink character reader

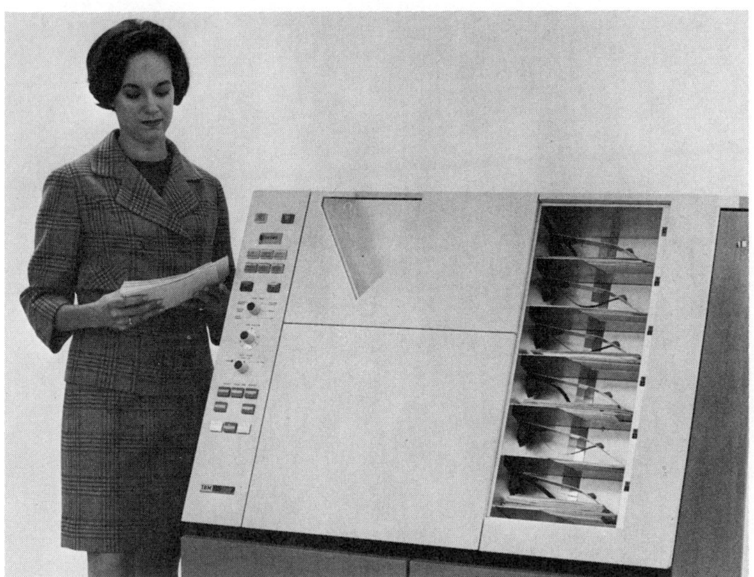

Courtesy of IBM.

**Figure 4.21.** Deposit slip

LOPES TRUST
OF JENSEY

DEPOSIT TO
ACCOUNT OF DATE ______________

FREDERICK OR MARY A LAHD

ADDRESS ______________________________
(PLEASE PRINT)
______________________________
CITY STATE ZIP CODE

ALL ITEMS SUBJECT TO VERIFICATION, COLLECTION AND CONDITIONS CONTAINED ON SIGNATURE CARD.

| ITEM COUNT | | DOLLARS | CENTS |
|---|---|---|---|
| | CASH | | |
| CHECKS | (list separately) 1 | | |
| | 2 | | |
| | 3 | | |
| | 4 | | |
| | 5 | | |
| | 6 | | |
| | 7 | | |
| | 8 | | |
| | TOTAL | | |

319 02400 8

to the CPU. New devices can be added when necessary, giving the computer system virtually unlimited storage capacity.

Output system. A variety of data output methods are available on the computer. In this step, the processed data is converted into a form readable by humans or suitable for further processing.

1. Line printer. The line printer shown in Figure 4.22 is a common output device. Line printers convert pulses from the CPU into characters

**Figure 4.22.** Line printer

Courtesy of Xerox Corp.

that are printed on a page. These machines can print up to several thousand lines per minute. Line printers are used to prepare documents, forms, and reports, called "hard copy" output.

2. Print units. The print unit on the console typewriter or a remote or local terminal is sometimes used to output small amounts of data. These print units produce hard copies of the data at a speed of 10 to 30 or more characters per second.

3. Punched cards and paper tape. These two are frequently used as output media for computerized systems. The results of computer processing are converted into the appropriate code and punched into cards or paper tape. The punched cards or tape can later be used as input to the computer system, or entered into a card or paper tape reader to produce a printed hard copy of the data.

4. Cathode ray tube (CRT). The cathode ray tube, or video display screen, displays an image called a "soft copy" of the data. The data is displayed as alphanumeric characters or graphics on a screen similar to the image on a television set. This method is used when a permanent copy

of the output is not required, such as in inquiry systems, table look-ups, and file search routines.

5. Audio response unit. This media also produces a soft copy. The device outputs results in the form of prerecorded spoken words. Audio response units allow a user to call a computer on the telephone and verbally receive the results of an inquiry or processing. The device selects the appropriate words and relays them over the telephone.

COMPUTER SOFTWARE. Computer software is the languages, programs, and instructions used to communicate with the machine and direct its operation. Many languages are available, with varying purposes, features, advantages, and limitations.

1. COBOL—COmmon Business Oriented Language. This is one of the most widely used languages in business and has excellent alphabetic and text manipulating ability. COBOL instructions closely resemble ordinary business English, allowing both professional programmers and non-programmers to trace the program logic easily. A limiting factor is the relatively large computer required to run the language.

2. FORTRAN—FORmula TRANslating System. This language is used mainly in the scientific and mathematic disciplines, as well as in business programming. FORTRAN instructions closely resemble algebraic notations, requiring documentation for ease of tracing program logic. FORTRAN is probably the most widely used computer language, and programs written in FORTRAN can be run on most small computers. It has extensive mathematical capabilities, but a somewhat lesser ability to manipulate text matter.

3. BASIC—Beginners All-Purpose Symbolic Instruction Code. This language, developed at Dartmouth College, is easy to learn and use. The instructions closely resemble ordinary English statements. BASIC was designed primarily for time sharing and interactive programming, is available on many small computers, and is offered by most time share firms. BASIC is widely used for business programming, particularly when time sharing facilities are involved.

4. RPG—Report Program Generator. This language is designed to facilitate the preparation of reports, particularly those which involve processing data files. RPG follows a fixed order of execution, input, processing, and output. It is simple to use, requiring only that the record layout follow a specific format. RPG is available on many small and large size systems and is widely used in business programming.

5. Assembler Language. This language is a compact, efficient system of abbreviations called mnemonic codes, which specify the operations that the machine is to carry out. Each brand of computer has its own Assembler language, and a program written for one machine will probably not run on other systems without some modifications. Assembler language is more

difficult to learn and use, but is the most efficient means of utilizing the primary storage area available on a computer when a program is being executed. For this reason it is used for coding many repetitive or lengthy business applications programs, as well as system software. On many small machines, it is the only language available.

6. Other languages. A variety of other languages, including PL/I, APL, and Algol, are also used for business programming. Since they require special computers or very large systems, they are not as widely used as those described above.

### Applications

Electronic data processing has its greatest value where large volumes of data are to be processed, and when speed and accuracy are important. Computers are particularly efficient at meeting these criteria, surpassing by far the capabilities of the unit record and manual methods.

Specific computer applications are varied. They include preparation of payroll, accounting forms, financial statements, billing and invoicing, inventory, stockkeeping and management reports. The mathematical capabilities of the computer finds use in financial analyses, interest and loan calculations, statistical applications, and other areas.

### Advantages

Computerized data processing has several major advantages. It can achieve a speed and level of accuracy far surpassing any other devices yet conceived. The computer excels in performing repetitive tasks with no loss of accuracy or speed.

The costs for processing large amounts of data are relatively low. It can conveniently perform a wide variety of procedures on data, including complex mathematical operations. Its ability to utilize a wide variety of input and output media gives the user greatly increased flexibility in data input alternatives and output formats.

Computerized data processing systems are ideal for storing large amounts of data. High-density storage media, such as magnetic tape or magnetic disk, enable millions of records to be stored in a small space, and at a relatively low cost. It would take several hundred thousand punched cards to hold the content of one $20 reel of magnetic tape, weighing about five pounds.

The stored program feature guarantees standardization of processing procedures to a greater degree than found in the unit record method. The ability of the computer to execute the stored program independently of human direction also encourages accuracy and standardization.

The remote processing capability of a computer system is a definite advantage. It expands the usefulness of a system geographically, as well as making inquiry and time sharing systems feasible.

## Limitations

Many of the limitations of computerized processing are in relative, rather than absolute, terms. More skills are required for writing programs and operating equipment than the level required for many activities in the unit record and manual methods. The cost of leasing or buying modern computer systems can involve many thousands of dollars, added to the cost of installing special air conditioning equipment or special floors.

One important limitation is the cost in time and money for processing a small volume of data, or "one time only" problems. In these instances, it may often be best to work the problem manually, using a desk calculator, or with unit record machines.

Another major limitation is that making modifications in equipment or procedures can be more costly in a computerized system than in other methods. New equipment may have to be purchased or new programs written and tested.

## EXERCISES

1. What are the major steps in the data cycle? Describe the function of each.
2. List at least five common data processing activities.
3. Describe the manual method of data processing.
4. What are the advantages and limitations of the manual method of data processing?
5. Describe the unit record method of data processing.
6. Select one of the unit record machines discussed in this chapter and describe its function.
7. Describe the advantages and limitations of the unit record method of data processing.
8. Describe the computerized method of data processing. List the major blocks and subsystems.
9. What is the function of the CPU?
10. Describe the stored program capability of a computer.
11. How does the input system differ from the output system?
12. Summarize the advantages and limitations of computerized data processing.
13. Interview a business employee and observe his or her task. Determine which method of data processing is used. Is it the best method?
14. Select a business firm which uses unit record processing, and describe the system, machines, and procedures used.
15. Select a business firm which uses electronic data processing, and describe the system, machines, and procedures used.

# Chapter 5

# Elements of System Input, Storage, and Output

One of the major considerations of the systems analyst is to plan and select the elements involved in the input, storage, and output functions of a system. Factors such as cost, retrieval time, permanence, file security, physical space considerations, method of recording original data, and types of output needed directly affect this selective process.

## ELEMENTS OF DATA INPUT

### The Source Document

A source document is a record of an original transaction. It is the paperwork generated at the time a transaction takes place. Examples include time cards prepared throughout the week by an employee; a list of expense account items gathered by a salesman as he travels; withdrawal slips prepared by a teller for a customer; or sales slips filled out by a salesperson at the time an item is purchased.

Source documents contain the data or information necessary for the procedures or activities initiated by the transaction. Time cards contain information needed to generate paychecks and entries in employment records. Withdrawal slips record information necessary for posting debits to customers' accounts. Sales slips contain the data that initiates activities in the inventory, stock room, delivery, personnel, or accounts receivable departments.

The data on source documents must be made available for processing by the system. Often data is recorded on source documents in longhand, and cannot be processed or read directly by machine. In these instances, the data must be converted into a form acceptable to the machines in use. This involves transferring the data on the source document to another document that is machine readable. This may be done by keypunching the data into punched cards, keyboarding it onto magnetic or paper tape, or by some other method.

Other forms of source documents are readable by both people and machines. The OCR devices, for example, eliminate the need to convert a source document into a machine readable form.

### Verification

Whenever data is transferred from a source document to another media by a human operator, the possibility of errors arises and the data must be verified. Verification involves checking the accuracy of the data transferred from one document to another by comparison. It is an essential step in data entry because it assures that no errors were made during the process of converting the source data.

Verification is accomplished in several ways, depending on the media used. When converting data to punched cards, a verifying machine such as the one shown in Figure 5.1 is used. First an operator punches the source data into cards on a keypunch. Then the cards are loaded into the verifying machine. The operator re-keyboards the same data from the

**Figure 5.1.** 059 card verifier

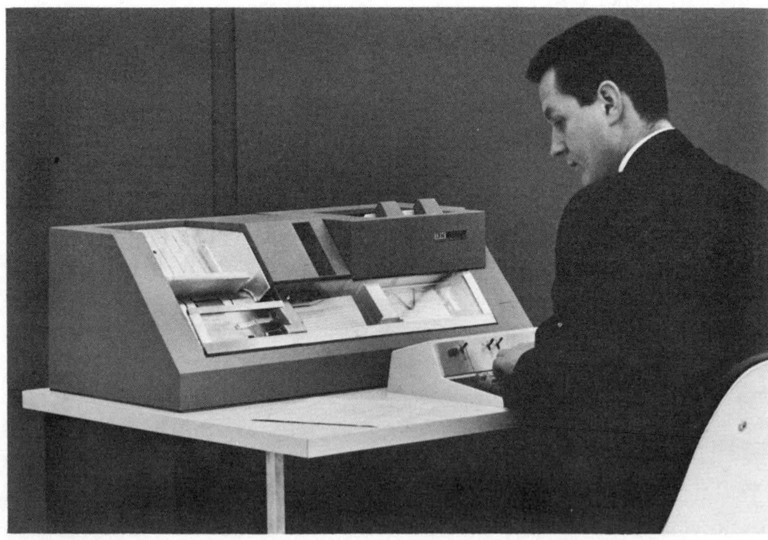

Courtesy of IBM.

source documents, and the verifying machine compares the keystrokes with those punched into the cards. If a difference is detected, the machine stops and notches the column on the card with the error. The operator determines the cause of the error—a misread figure, keyboarding error, machine malfunction—and keypunches a corrected card.

Verified cards are assumed to be free of transcribing errors. Of course, errors in the source documents (such as an incorrect number of hours on a time card) would go undetected.

Some keypunches and machines that record data on magnetic tape or disk use a different method. The data keyboarded from source documents is held in temporary storage in the machine while the operator re-keyboards the same data. If the two sets of data agree, the machine completes the transcription and records the converted data on the input media. If they do not agree, the operator stops, determines the cause of the error, corrects it, and goes on to the next record.

Another method of verifying transcribed data is to use batch totals. After a group of records have been converted from one media to another, a batch total of the transcribed figures is drawn and compared to a batch total of the original figures from the source documents.

### Offline and Online Data Input

Data may be input to a processing system in one of two modes: offline and online. Offline data entry involves preparing data to be input to a system at a later time. In this mode, source data is transcribed onto a machine readable storage media, such as paper tape, magnetic tape, or punched cards. The stored data is held in this form until needed for processing, at which time it is input to the system.

In online data entry, the source data is entered directly into the processing system at the moment the transaction occurs, without any intermediate storage. Systems using online data entry are nearly always computerized. Source data may be keyboarded directly onto a terminal which is connected online to the computer. Input is real time . . . the data is entered for processing at the time the transaction occurs.

No conversion of data from one mode to another takes place. The data entered is usually displayed or printed in a form the operator can read, and at the same time transmitted in machine readable form to the computer. Examples include bank teller terminals and reservation inquiry systems.

Media such as optical character recognition readers and magnetic ink character recognition systems, which can be read by both people and machines, are used for both on- and offline input.

### Input Media

Several different methods and machines are used to record and prepare source data for input into the system. The most common are described below:

1. MANUAL DATA ENTRY. The least sophisticated and most prevalent method is the manual entry of data onto source documents, done by longhand or with a typewriter. The source documents are often suitable for further processing in the manual method and can be input to the system without additional conversion or copying of data.

Paychecks are prepared from handwritten time cards. Invoices are prepared from sales slips. Orders are picked and delivered using information recorded in longhand on original orders forms. Data captured manually must usually be converted into machine readable form for further processing by other methods.

2. PUNCHED CARDS. Punched cards are a widely used input media for both unit record and computer systems. The standard 80-column card measures 3¼″ × 7⅜″ and holds up to 80 alphabetic, numeric, or special characters. A combination of one or more holes punched into each column represents each character. The characters punched into the columns are usually interpreted or printed out along the top of the card. The coding system used to encode information on the card is shown in Figure 4.3. Numbers require one punch; alphabetic characters have two.

The 80-column card is divided into twelve rows. The two at the top are unlabelled and are referred to as the 11 and 12 zones. The other 10 rows are usually labelled 0 to 9 on the face of the card. (The zero row is sometimes also considered a zone.)

Figure 4.5 shows the 96-column card. This card is the principal means of data entry for the IBM System/3 computer. The card measures 2⅝″ × 3¼″ and contains 96 columns of data punched into three tiers, or levels. The upper portion of the card contains space for interpreting or printing up to 128 characters. The lower portion contains the punched information.

Information recorded in punched cards often occupies several adjacent columns. This group of related columns is called a field. For example, a description of machine parts with 20 characters would occupy a 20-column field. A two-character identification number would occupy a two-character field. Data keypunched into cards from source documents must be verified to assure accuracy.

3. MARK SENSE CARDS. The mark sense card shown in Figure 5.2 is another method of data entry. Data is encoded on a card by marking in the appropriate bubbles with an electrographic pencil. The cards are then run through a machine such as the IBM 519 Document Originating Machine, which converts the marked bubbles into punched holes in the same cards. After punching, the cards are machine readable and ready for further processing by a unit record or computer system.

An advantage of mark sense cards is that source data may be prepared for system input with only a pencil. No special punches or machines are required. This facilitates preparation of input data in the field, eliminates the time and effort required for data conversion, and increases the level of

**Figure 5.2.** Mark sense cards

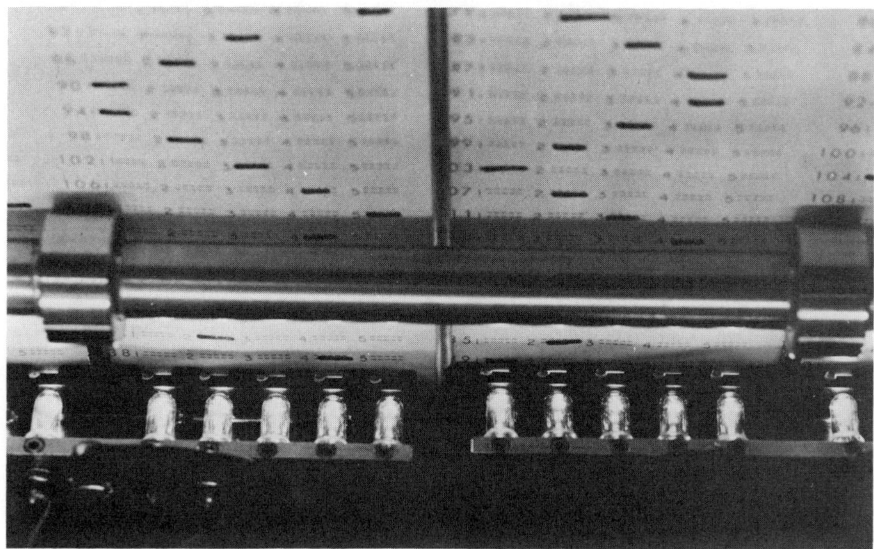

Courtesy of IBM.

accuracy. Since no conversion of data between source document and input media is involved, data need not be verified.

An example of mark sense usage would be a clerk taking a physical inventory and recording the count directly on mark sense cards. The cards are later converted to machine readable form without any keyboarding.

4. PORT-A-PUNCH CARDS. A pre-punched, diecut card may be used for data entry. Source data is entered by punching out the appropriate holes in a card with a stylus and a portable board (Figure 5.3). Since the punched data is in a machine readable form, no further data conversion or verification is necessary. Portable punch boards and cards provide a quick, convenient means of recording source data suitable for machine processing while in the field.

Voter tally systems often use this means of data entry. It is economical and fast, and data can be processed without any further conversion.

5. OPTICAL RECOGNITION DEVICES. Optical recognition records source data in a form that is readable by both people and machines. Some machines read only filled-in areas; others can distinguish the letters of the alphabet and numeric digits by their optical characteristics.

There are two principal means of recording data:

a. Bar Marks. In this system, an optical device senses a filled-in area. It may be bubble, a space between two lines, or a slash through a printed character. The device converts it to the appropriate electrical

**Figure 5.3.** Port-a-Punch

Courtesy of IBM.

pulse for entry to the computer system. Printed characters on the face of the document identify the data being encoded to make it readable by humans. This type of document is commonly used for such things as recording meter readings, inventory control, and test scoring. (See Figure 5.4.)

b. Optical Character Recognition (OCR). Some input machines, such as the IBM 1288 Optical Character Reader, are designed to recog-

**Figure 5.4.** OCR bar marks

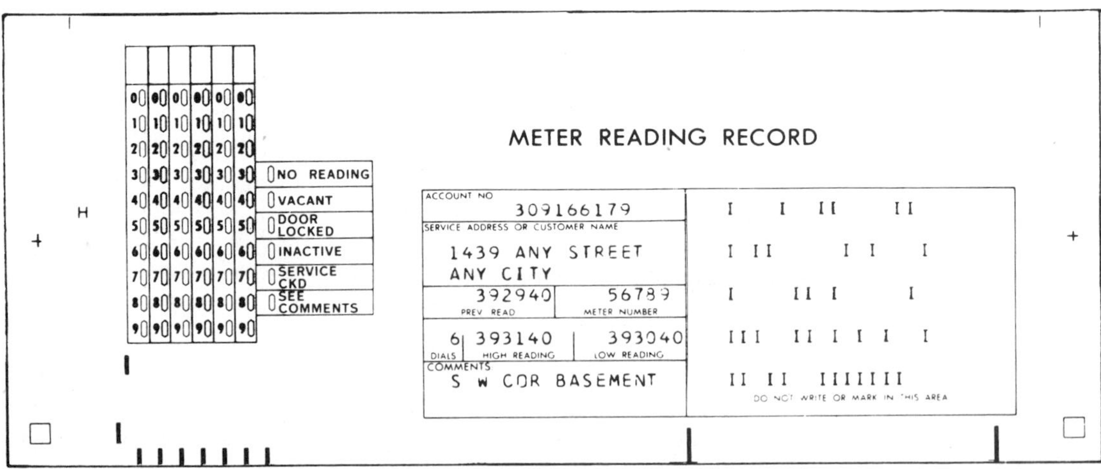
METER READING RECORD

NO READING
VACANT
DOOR LOCKED
INACTIVE
SERVICE CKD
SEE COMMENTS

ACCOUNT NO
309166179
SERVICE ADDRESS OR CUSTOMER NAME
1439 ANY STREET
ANY CITY
392940 PREV READ
56789 METER NUMBER
6 DIALS
393140 HIGH READING
393040 LOW READING
COMMENTS
S W COR BASEMENT

DO NOT WRITE OR MARK IN THIS AREA

Courtesy of IBM.

nize the shape of a letter form. Some models will read clearly printed handwritten numbers and control characters. Others read certain fonts of typewritten or printed text. The device converts these images into electrical pulses suitable for computer processing.

In optical character recognition, the image area to be read is divided into many sections. The shape of each character or letter fills in different sections, forming a unique pattern. The OCR device uses an optical scanner to trace the form of the character, and converts it to a corresponding pattern of pulses which are sent to the computer for processing.

Examples of alphameric characters used in this type of data entry are shown in Figure 5.5. This form of data entry is used to prepare such things as sales slips, bills, personnel information, and order forms.

OCR input is a very flexible means of data entry. Its chief advantage is that the source document can be read by both people and machines, and it is usable by people for further processing. There is no need for data conversion and verification. OCR documents are also suitable for both online and offline data entry.

6. MAGNETIC INK CHARACTER RECOGNITION (MICR). Magnetic ink character recognition is another means of preparing source data for direct entry into a computer system. Characters to be read are printed on a page using a special magnetic ink. When these printed letters are passed through a magnetic field, they change the characteristics of the flux, and the device sends a string of pulses, representing a given character, to the computer. No conversion or verification of data is necessary since the source document is used for the data entry to the system.

MICR is widely used in the banking industry to encode account num-

**Figure 5.5.** OCR input characters

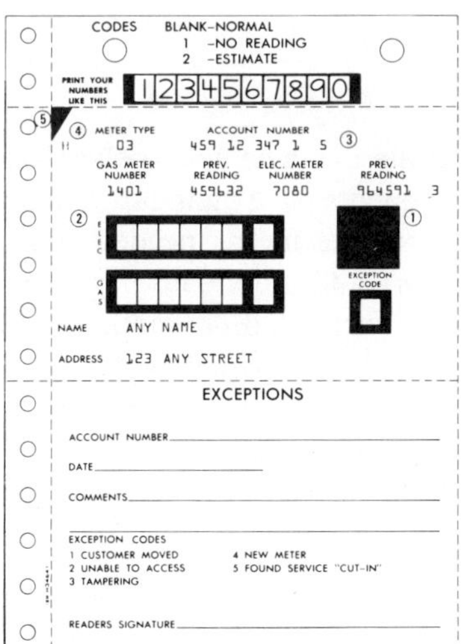

CODES BLANK–NORMAL
1 –NO READING
2 –ESTIMATE

PRINT YOUR NUMBERS LIKE THIS 1 2 3 4 5 6 7 8 9 0

(5) (4) METER TYPE 03 ACCOUNT NUMBER 459 12 347 1 5 (3)

| GAS METER NUMBER | PREV. READING | ELEC. METER NUMBER | PREV. READING |
|---|---|---|---|
| 1401 | 459632 | 7080 | 964591 3 |

(2) ELEC (1)

GAS

EXCEPTION CODE

NAME ANY NAME

ADDRESS 123 ANY STREET

EXCEPTIONS

ACCOUNT NUMBER________

DATE________

COMMENTS________

EXCEPTION CODES
1 CUSTOMER MOVED
2 UNABLE TO ACCESS
3 TAMPERING
4 NEW METER
5 FOUND SERVICE "CUT-IN"

READERS SIGNATURE________

Courtesy of IBM.

bers and amounts on checks and deposit slips. The American Banking Association has approved a standard letter shape and character position for encoding data on checks and forms. The recorded data may be read both by an MICR reader and the human eye.

7. KEYBOARD TERMINALS. A keyboard terminal connected to a computer is another method of entering source data into a computer system. As the operator depresses each key, electronic pulses representing the character are sent to a computer. At the same time, a hard copy of the data is printed out on the terminal for use by people. This method of data entry is online and real time.

The keyboard terminal might be the console typewriter connected to the computer and used to input small amounts of data for a program. The more common arrangement would be for the keyboard terminal to be at a remote location—in another room or in another part of the city.

This method is widely used for entering data on savings and loan transactions. An online teller terminal is placed near each teller. The teller enters the data regarding a transaction as it is made. As each key is struck, the data is held in a holding device called a buffer. When the line is visually verified by the teller as correct, a key is depressed and the entire line is transmitted to the computer for immediate processing. A hard copy is sometimes printed out on the teller terminal for use in posting, batch totals, and the like.

## DATA STORAGE MEDIA

Following are the media most commonly used for storing data. Each system has its advantages and limitations. The method selected by the systems designer will depend upon cost, permanence, accessibility, and security, as well as other factors.

### Manual Storage Media

1. VERTICAL FILES (Figure 5.6). Vertical file cabinets and shelves are suitable for storing originals or copies of documents in a manual system. Documents can be filed easily and removed from the file as needed. Files can be searched manually, either sequentially or randomly. The degree of accessibility depends on where the cabinets are located. File cabinets are relatively fireproof, and additional security can be provided with locks, limited access rooms, and specially built cabinets. Vertical files are widely used for data storage because of their low cost and availability. They are, however, bulky and inefficient for holding large volumes of data, and they require manual retrieval of data.

2. VISIBLE FILES (Figure 5.7). Visible files are similar to vertical files except that records are stored in a horizontal position, partially overlaid by

**Figure 5.6.** Vertical file

**Figure 5.7.** Visible files

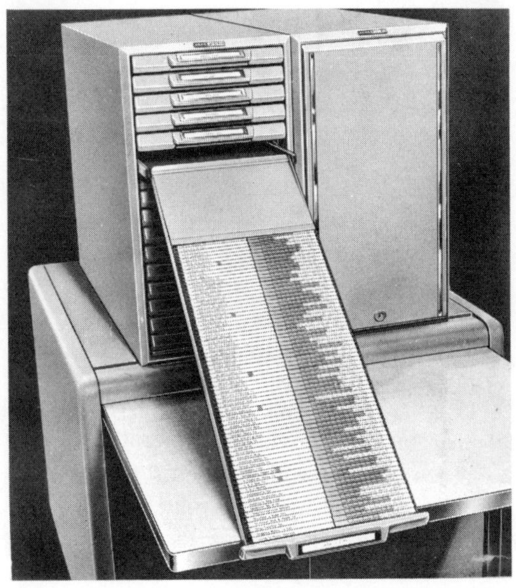

the next record, so that a portion of each record is visible. The arrangement resembles the way shingles are laid on a roof. Visible files are used to store originals or copies of documents. They are searched manually, either randomly or sequentially. Accessibility depends on the location of the files and the amount of data stored in them. Due to their design, visible files can be accessed faster than can vertical files, but they are too large to be efficient for storing large volumes of data.

3. OPEN FILES AND TUB FILES (Figure 5.8). Large tubs or open trays of files are often used to store records that must be readily accessible to an operator. Their open design enables an operator to reach over conveniently and replace or pull a record from the files without leaving his or her seat. These files, however, occupy too much floor space to be efficient for storing large quantities of data.

**Figure 5.8.** Tub file

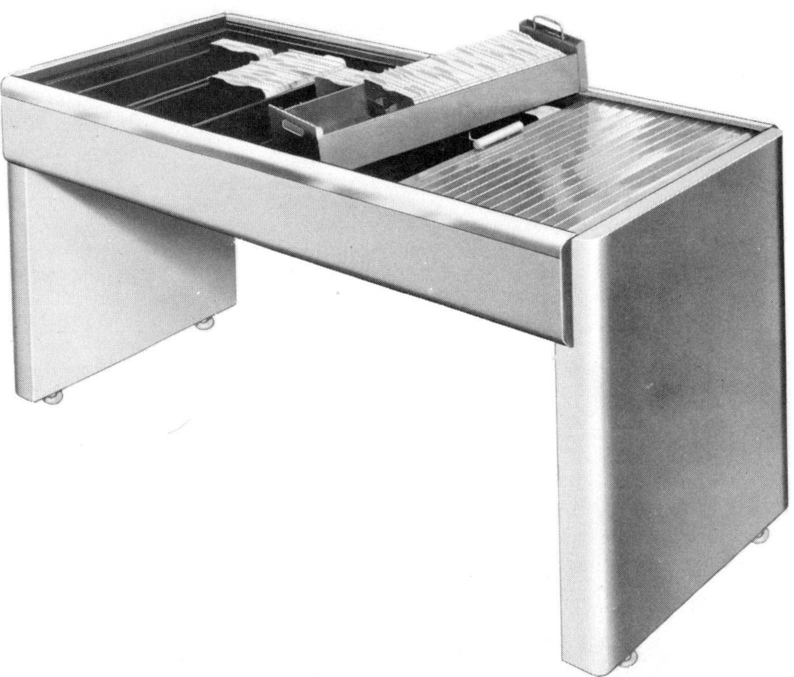

4. ROTARY AND WHEEL FILES (Figure 5.9). Records are often placed in carrousels or attached to wheels for faster access. The operator rotates the unit until the desired record is located. On some designs the records are permanently attached to the rotor, preventing misfiling or loss. This type of file is, however, limited in the number of records that may be maintained on a single rotor.

5. AUTOMATED FILES (Figure 5.10). Various forms of automated files are used to store information and records for manual retrieval. An operator directs the device to rotate to a specific section of the file, locates the required record, and removes it manually. Automated files are used where a large volume of source records must be maintained for quick access. They save time and steps in locating a record, but are more expensive than other manual systems.

**Figure 5.9.** Rotary files

Courtesy of Acme Visible Records, Inc.

**Figure 5.10.** Automated files

Courtesy of Acme Visible Records, Inc.

## Unit Record Storage Media

Unit record storage systems are designed to hold 96- or 80-column punched cards. Figure 5.11 illustrates one such device. The drawers of punched cards are usually placed in vertical file cabinets. The cards can be searched manually (if they are interpreted) or carried to a piece of unit record equipment and searched automatically. Because of the inefficiency of storing data on punched cards, unit record systems are not usually used to hold large volumes of data.

**Figure 5.11.** Unit record storage

Courtesy of Art Steel Co., Inc.

## Computer Storage Media

Electronic storage media are the most widely used means of storing large volumes of numeric or alphabetic data. These systems store data that has been converted from source documents into machine readable form. The stored data is subject to accidental erasure, and measures must be taken to assure file security. The accessibility of the data stored on these media is indirect. The stored data must often be converted, by computer or peripheral device, back into printed characters before they can be used by humans. The following are the principle means of computer storage:

1. MAGNETIC TAPE STORAGE. Data may be stored on reels of magnetic tape, which is a plastic ribbon coated with a ferromagnetic material. It measures ½″ wide and is wound on reels in lengths of 1200″ and 2400″.

(See Figure 5.12.) Each inch of tape holds from 560 to 1600 bits of information, depending on the particular system. A single reel of tape holds about 40 million characters.

**Figure 5.12.** Magnetic tape

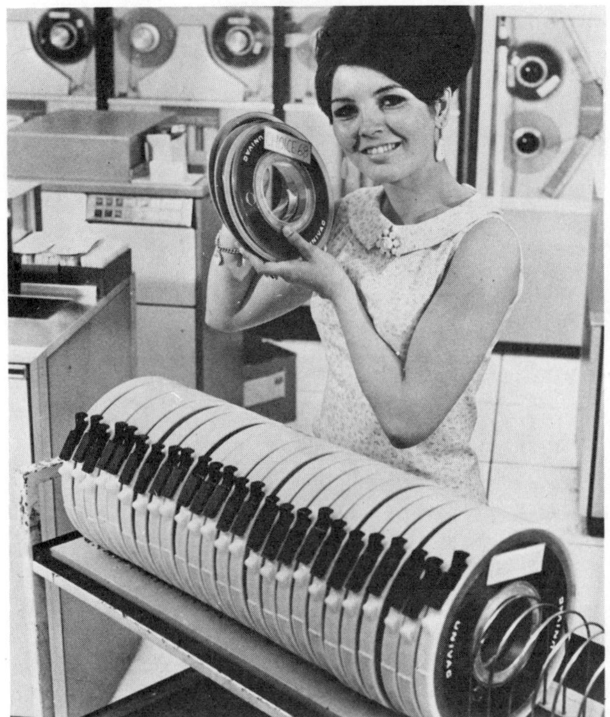

Courtesy of Sperry Univac, a division of Sperry Rand Corporation.

The reel of tape is placed on a tape drive shown in Figure 5.13. Read/write heads on the tape drive encode information on the tape or read the data already recorded. Because of its nature, recording tape is a sequential access media. Each record on the tape must be searched in sequence until the desired record is located. The average access time of magnetic tape is five seconds, slower than most other computer storage media.

2. MAGNETIC DISK STORAGE. The magnetic disk system (Figure 5.14) is a fast access media for storing data for computer processing. Magnetic disks are coated with a ferromagnetic substance, capable of being magnetized to represent binary data. Individual disks are grouped into disk packs and mounted on drives. (See Figure 5.15.) Read/write heads located between each of the disks access the top side of one disk and the bottom of another. Data is read or recorded by the heads in concentric circles on the disks, as they are rotated by the drives. Since the read/write heads can move independently to any point on the disks, random access of data is

**Figure 5.13.** Magnetic tape unit

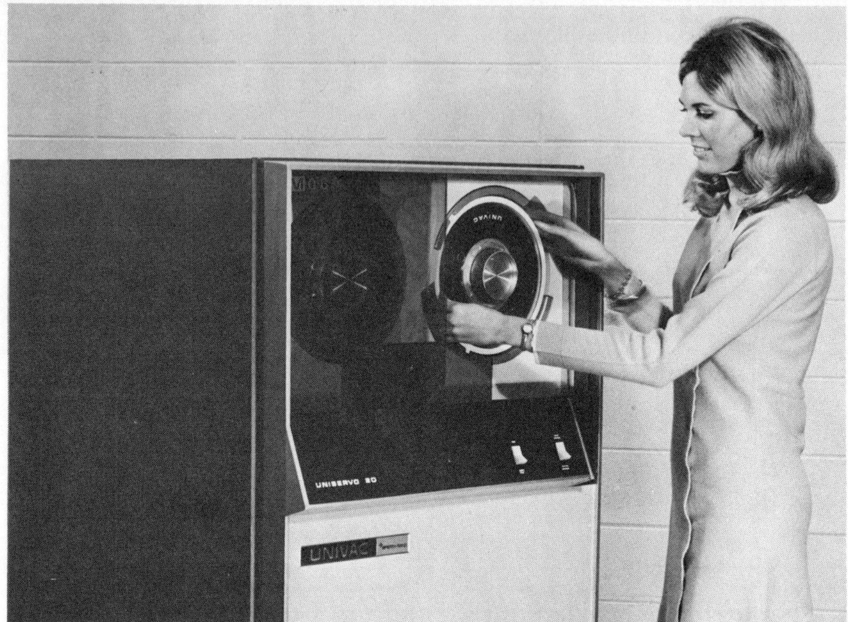


Courtesy of Sperry Univac, a division of Sperry Rand Corporation.

**Figure 5.14.** Disk pack

Courtesy of Sperry Univac, a division of Sperry Rand Corporation.

possible. The average access time on a disk storage system is 75 milliseconds. About 30 million characters can be recorded on a single disk pack. The packs can be removed from the drive and stored in a cabinet.

3. MAGNETIC DRUM STORAGE. Computer systems may be equipped with magnetic drum storage devices. (See Figure 5.16.) These devices record data on a revolving drum that has been coated with a ferromagnetic

**Figure 5.15.** Disk storage drive

Courtesy of Hewlett-Packard.

**Figure 5.16.** 2303 drum storage device

Courtesy of IBM.

material. The drum passes under read/write heads that read information or encode it on the magnetic surface.

The average access time of magnetic drum storage devices is 8.6 milliseconds. Drum devices will store up to 4 million characters per drum. This media is used where speed and random access of data are important.

4. DATA CELL STORAGE (Figure 5.17). This device is composed of a large cylinder, called a data cell drive, which holds a group of ten cells. Each cell has 20 subcells, composed of 10 data strips. Each strip is coated with a ferromagnetic material and holds up to 100 tracks of recorded data.

In operation, the system pulls the appropriate data strip from the device and wraps it around a drum. A read/write head located over the drum reads from, or records data onto, the strip. Then the strip is replaced in its proper place in the drive.

Data cell systems are a random access media with an average access time of 350 milliseconds. About 400 million characters can be recorded on a single drive, making this an efficient means of holding a very large volume of data with a relatively fast access time.

**Figure 5.17.** 2321 data cell drive

Courtesy of IBM.

## Photographic Storage Media

Photographic imaging techniques for storing data offer several important advantages. Photographic representations of data are much smaller in size than the original document, saving storage space. Photographic techniques allow signatures, pictures, drawings, and other source material and artwork, as well as text, to be copied and stored.

The photographed representation of the data is a fairly permanent means of storage, and it meets many legal requirements. Data is recorded in a form readable by people, but a mechanism that enlarges the image is usually required.

Accessibility can be sequential or random. The degree of accessibility and convenience depends on the design of the particular system and the location of the stored files.

Microform storage is the term given to photographic reduction of data for storage and retrieval. An entire source document may be reduced as much as 50:1—a drawing, 45″ × 63″, becomes the size of a punched card, or smaller.

MICROFORM STORAGE. The microform process involves four steps, shown in Figure 5.18. Microfilming the source document is the first step. A microfilm camera is used to copy the document onto film. Up to 200 documents (8½″ × 11″) or 640 checks can be photographed per minute.

**Figure 5.18.** Microfilm process

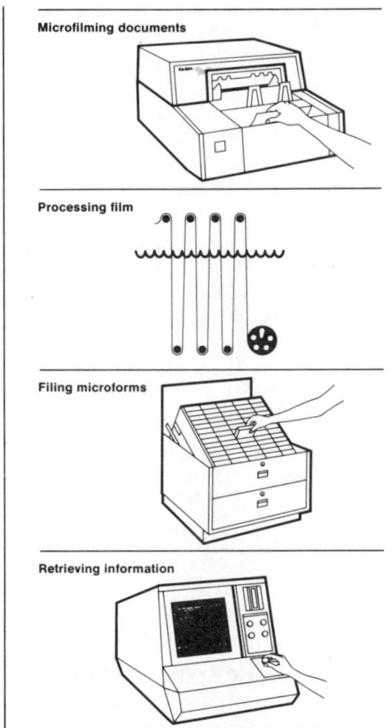


Courtesy of the Eastman Kodak Company.

The second step is processing. Here a negative or positive film image, many times smaller than the original document, is developed.

The third step involves filing the microform. This may involve mounting the film into special carriers, or cartridges, and placing them in some form of filing system.

The last step is retrieving the information from storage. The appropriate microform record is pulled from storage and displayed on a microform reader or viewer.

ADVANTAGES OF MICROFORM STORAGE. Microform techniques possess several advantages; for instance, a significant savings in storage space. Microform records occupy as much as 98 percent less space than the original documents. And since storage space costs money, this usually results in a financial savings as well. Filing costs are often reduced also, since a smaller size document is being filed and retrieved. It is more economical to distribute microform copies of a file than to duplicate and distribute the original file by conventional printing means. Mailing charges are less. Microfilmed records are often admissible as primary evidence in a court of law in instances where other storage media are not.

MICROFORMS. Microform storage is available in five common forms, all of which use photographic imaging techniques.

1. Roll Microfilm (Figure 5.19). Original documents and records are copied onto strips of film, measuring 16mm or 35mm wide. The film is wound on reels for ease of handling.

**Figure 5.19.** Microfilm roll

Courtesy of the Eastman Kodak Company.

Since the resulting reduced image is too small to be viewed adequately by the naked eye, the reel is placed in a microfilm reader, shown in Figure 5.20, for viewing. This device is equipped with a projection mechanism

Figure 5.20. Microfilm reader

Courtesy of the Eastman Kodak Company.

which enlarges the image many times. Reels of microfilm records may be easily filed or duplicated.

2. Microfilm Magazine (Figure 5.21). In this method, the reel of microfilm is enclosed in a plastic magazine, measuring 4″ × 4″ × 1″. The magazine adds convenience and flexibility to the system, can be easily filed, and protects the film as well. Magazines are used with self-threading viewers.

Figure 5.21. Microfilm magazines

Courtesy of the Eastman Kodak Company.

3. Microfilm Jacket (Figure 5.22). A processed length of microfilm may be cut into strips and mounted in a plastic jacket, providing a convenient method of handling related microfilm images. The standard jacket is available in sizes from 3″ × 5″ to 5″ × 8″, and is easily mailed, filed, or sorted.

**Figure 5.22.** Microfilm jacket

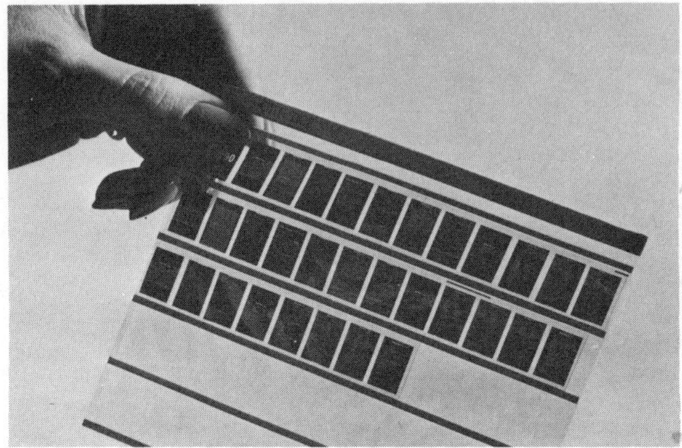

Courtesy of the Eastman Kodak Company.

4. Microfiche (Figure 5.23). The microfiche is a popular form of microfilm. A group of microfilmed images, usually 98, are exposed on a single piece of film measuring 4″ × 6″. Sometimes greater reductions enable up to 500 images to be filmed on a single card. To view one of these micro records, the card is placed in a microfiche reader, shown in Figure 5.24.

**Figure 5.23.** Microfiche card

Courtesy of the Eastman Kodak Company.

Figure 5.24. Microfiche reader

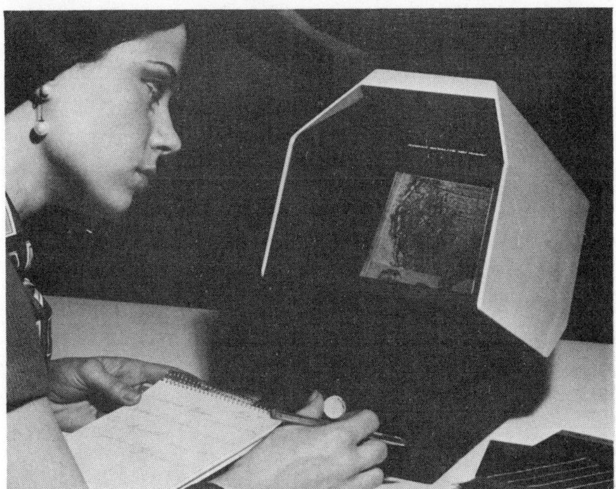

Courtesy of the Eastman Kodak Company.

The reader is indexed to the desired image and projects it on a viewing screen.

Microfiche cards are conveniently copied, duplicated, or mailed. Catalogues, reports, and other multipage documents are often recorded on microfiche.

5. Aperture Card. A standard 80-column card is often used to hold either 16mm or 35mm microfilmed images. (See Figure 5.25.) The film is mounted in a window in the card, called an aperture. A reader is used to enlarge the size of the image for convenient viewing.

Figure 5.25. Aperture card

Courtesy of the Eastman Kodak Company.

The advantage of aperture cards is that they can be processed by conventional unit record machines. Large engineering drawings, artwork, and pages of text, for example, can be sorted, selected, merged, or filed.

## OUTPUT MEDIA

A variety of output media is available to report the results of data processing. They vary in cost, permanence, and function. The media can be combined in different ways to produce a variety of output formats suited to the particular needs of an individual system. The major forms of output are:

### Hard Copy Output

Hard copy output consists of documents, forms, and reports typed or printed out on paper. This media is usually selected when the results of processing must be human readable, saved permanently, or handled by people. (See Figure 5.26.)

**Figure 5.26.** Hard copy output

Courtesy of Sperry Univac, a division of Sperry Rand Corporation.

The common forms of hard copy output are:

TYPEWRITER. Data output from the manual method is usually generated on a standard or electric typewriter. The speed at which the output is generated is limited by the ability of the operator and the type of equipment used. The typewriter is also sometimes used to generate output in the unit record system.

UNIT RECORD MACHINES. Hard copy output in the unit record system is often generated on devices such as the IBM 407 Accounting Machine and the UNIVAC 1004 III. These machines produce a printed copy of the output as well as punched cards suitable for further processing.

ONLINE TYPEWRITER. Console typewriters or remote or local terminals are often used as output media in computer data processing systems. These devices produce a typed hard copy of the results on paper. The rate of output ranges from 10 to 30 or more characters per second. Online typewriter output is suitable when a relatively small volume of data is to be printed out.

LINE PRINTER. Line printers are common output devices found in many computer systems. They are capable of producing hard copy output such as documents, reports, and printed forms. A line printer outputs all the characters in one line simultaneously at the rate of several thousand lines per minute. This makes them the most practical media for producing hard copy output of a large volume of data.

## Soft Copy Output

Soft copy output consists of a non-permanent or transistory output in the form of a video display or audio response. Such output is used in computer systems whenever a permanent record is not necessary. (See Figure 5.27.)

The most common soft copy media are:

VIDEO DISPLAY. A cathode ray tube (CRT) displays letters, lines, curves, or charts on a screen similar to a television set. Video display

**Figure 5.27.** Soft copy output

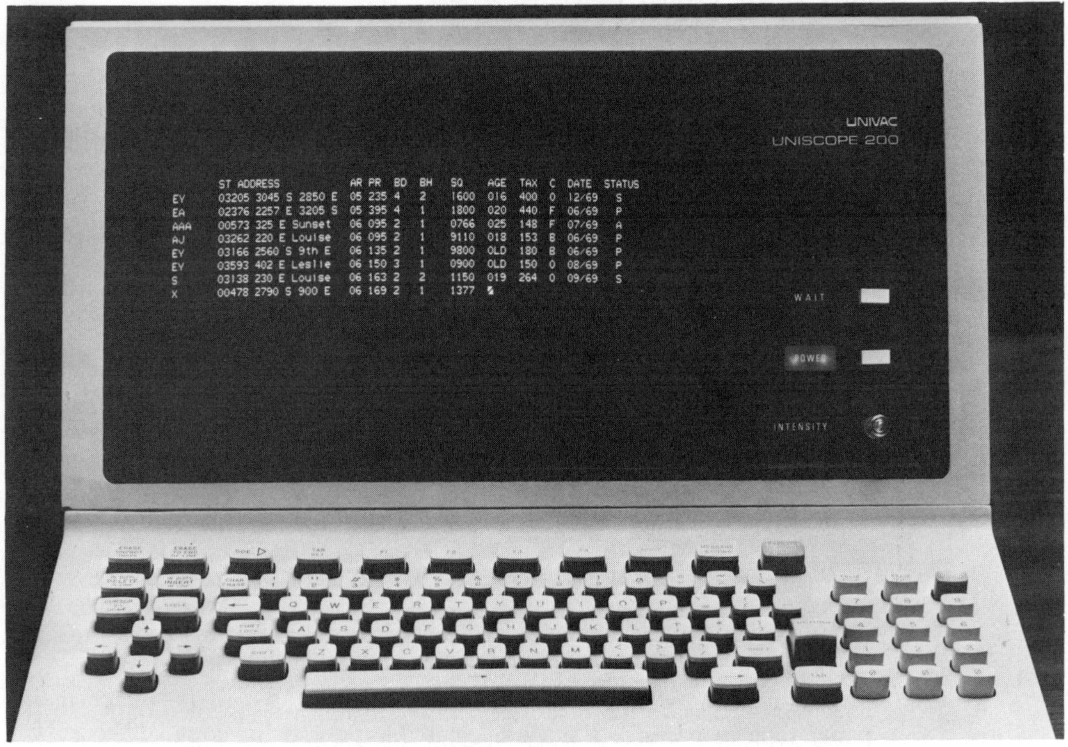


Courtesy of Sperry Univac, a division of Sperry Rand Corporation.

systems can output a large volume of data relatively rapidly, and are practical for use in query systems, searches, and for displaying data in storage.

AUDIO RESPONSE UNIT. Audio response units output the results of processing as a prerecorded spoken word. These units construct words or sentences from sounds that have been recorded on magnetic drums, and output them into a telephone system. Audio response units are used to output responses in query systems.

### Machine Copy Output

The third method of outputting data from a system is in machine readable form. This output is used to store results of processing when the data is to be reentered into the system for further processing. (See Figure 5.28.) Some forms of machine readable output can be read by humans as well.

**Figure 5.28.** Machine copy output

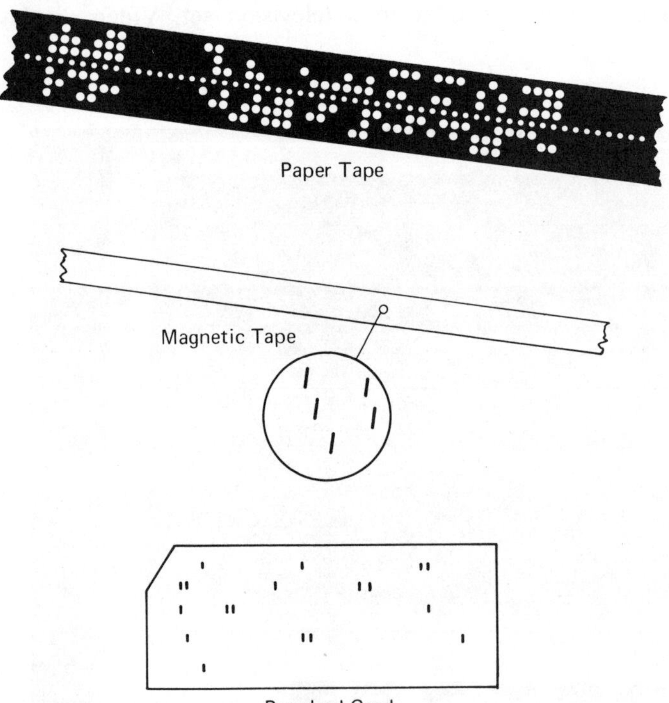


The most common forms of machine readable media are:

PAPER TAPE. Data represented by coded holes can be punched into paper tape to produce a machine readable permanent copy of the output. These tapes can be saved, read back into the system for further processing, or used offline with terminals to produce hard copy output.

MAGNETIC TAPE. Data can be output on magnetic tape for storage or further processing. Magnetic tape units can be used online to read the information back into the computer, or offline with other devices to produce other forms of output. Magnetic tape is an efficient and economical media to use when medium to large quantities of data are involved.

PUNCHED CARDS. Data can be punched into cards to produce machine readable output. The punched data can be input back into computer or unit record systems for further processing. Punched cards can be used offline with unit records devices to produce other forms of output.

### Output For People and Machines

Often data is output both in machine readable form for further processing and in human readable form during the same processing procedure. A common arrangement is to output processed data, such as updated accounts receivable files, on magnetic tape and print out monthly statements on the line printer at the same time.

A more sophisticated arrangement is *Computer Output Microfilm* (COM). Data, textual information, charts, and the like are recorded on magnetic tape by a computer, producing a copy that is efficient for storage and machine readable for further processing. This magnetic tape is run through a device that converts the electronic pulses to visual images on a video screen (CRT). The images are then photographed on microfilm, producing another copy of the data that is human readable and in a form that is easily stored. This arrangement has the additional advantage of producing large quantities of data output at relatively high speeds.

## SELECTION OF DATA STORAGE MEDIA

Data storage is the process of recording data and information on a media and saving it for later retrieval by a system. The type of storage system selected depends on the requirements of the company, the amount of data to be stored, and the type of access needed. The following variables should be considered when designing a storage system:

### Cost

Cost is always an important criterion when selecting a storage system, but it is particularly important when large volumes of data are to be stored and retrieved. The two major considerations in figuring costs are the space available for storage and the amount of data to be stored.

Space is very often an expensive commodity, and the space requirements for a particular storage media is an important factor in hardware selection. As a rule, it is very expensive to store large volumes of source documents in their original format. They require bulky file cabinets of some type, and often a considerable amount of floor space. Usually, relatively small amounts of data are recorded on each document.

Storing unit records are also bulky and costly if a large volume of data is to be saved. Only 80 to 96 characters may be recorded on a single record, and the cards are not reusable.

Magnetic tape, or some form of microfilm, is the most inexpensive means of storing a large volume of data. A reel of magnetic tape costing $20 can hold the equivalent of several hundred thousand punched cards. A file recorded on microfilm may require only 2 percent of the original space occupied by the paper source document.

The amount of data to be stored is another important consideration when selecting storage media. If only a small amount of data is involved, saving original documents in file cabinets may be convenient and adequate. Medium amounts of data, such as an accounts receivable file for a small merchandiser, are often stored on punched cards or magnetic tape. Large amounts of data, such as involved in saving research papers for a library or in large inventories, are best stored on a media such as microfilm or microfiche.

### Permanence

How long data must be saved is another important consideration in storage media selection. Some documents and records must be kept permanently on file for tax, legal, or audit purposes. Often original documents must be saved even though the data they contain has been transferred to another media. Other data may need to be saved only for short periods, until confirmation has been received or transactions completed.

The degree of permanency and protection varies with the storage media. Magnetic tape, for example, can be erased or degaussed (demagnetized). Punched cards can be torn, mutilated, or stapled. Data centers are subject to fire and vandalism. Errors in processing can occur, erasing old data or producing new files with erroneous data. If source documents have been saved, it is often possible to regenerate damaged files. In other instances, backup or duplicate files on microfilm may be made to preserve data in case of emergency.

### Legal Considerations

The legality or validity of the form in which data is saved by a storage media must be considered when planning a system. Copies of signatures, contracts, or other documents are not always acceptable—many times only the original document will suffice. In other circumstances, photographed records of source documents are legally acceptable.

In the insurance industry, for example, files of original signatures of policyholders must be maintained, even though all other data has been transcribed from source documents to machine processable records.

## Accessibility

Accessibility refers to the degree of difficulty involved in retrieving documents or data from storage. The most direct form of data accessibility, is, of course, to have documents or records filed in cabinets directly available to the person who needs them. This system, however, becomes somewhat cumbersome when very large quantities of data are involved. In these instances, records and documents are often sent to warehouses away from the business enterprise, where space is less of a premium and greater fire and physical protection can be maintained. Data is retrieved by having someone physically remove the documents from their storage locations in the warehouse and return them to the requesting office. This takes a certain amount of time, and the system must be carefully planned so that necessary data can be easily located when needed in the processing cycle.

Data stored electronically provides a less direct means of accessibility. The convenience afforded by this method depends on the arrangement of the particular system. When large amounts of data are involved, electronic storage is, in the long run, the most convenient and the fastest way to access data.

## Access Time

Access time is the time required to locate a given record in a file. Average access time is determined by the number of records to be searched and the speed of the retrieval system. It takes longer, of course, to search a large number of records than it does to search a smaller number. And a slow retrieval system will naturally take more time to locate a record than will a faster retrieval device.

Since the slowest means of accessing records is often the human hand, manual methods are time consuming, costly, and inefficient for searching large files. Computerized methods are much more practical here. Such devices are magnetic tape, magnetic disk, and aperture cards have much shorter access times and are more efficient for retrieving data in large files.

Accessibility time is also determined by the type of search performed. The most common types are sequential and random searches. In sequential searches, records are checked one after the other in order, until the correct one is located. Magnetic tape and unit record devices locate data using sequential searches.

Random access devices have the ability to go directly to the data stored in a specific location in a file without first checking other records. This consumes far less time than would a sequential search, especially if the file is large and the object of the search is not near the beginning.

Magnetic disk, drum, and data cell devices can perform random as well as sequential searches. The human operator also has the ability to perform both types of searches.

Generally, there is a trade-off between access time and cost—the faster the access time, the greater the storage cost. It will cost more money to store large volumes of records on magnetic disk, for example, than on magnetic tape. But the average access time is less for disk than for tape.

### Security

Another important consideration when selecting a storage system is the possible degree of security and document protection. Confidential information, such as prices, customer lists, or discount rates, are often part of data files and should be accessed only by certain people. Protecting and preserving this type of information is often essential to the survival of a given business firm.

Data storage systems should be designed with this need in mind. Confidential files should be placed in controlled areas, where they can be accessed only by authorized personnel. Special code numbers, passwords, and security checks are techniques used to maintain security of data in storage systems.

Files must be protected not only against unauthorized access, but against physical threats as well. Fire, earthquakes, or water damage threaten the security of data and documents. Backup or duplicate files are often maintained as security measures. In the event of loss or destruction of the master file, the data can be regenerated from the backup file.

## EXERCISES

1. Define "source document" and describe how it differs from other documents used in a business.
2. Define "verification" and explain why it is important.
3. Contrast the differences between offline and online data input.
4. List four different input media or methods.
5. How does OCR input differ from MICR input?
6. Describe the major types of computer storage media available.
7. What are the advantages of storing data on photographic media? List several examples where it might be used.
8. How do photographic storage media differ from electronic storage media?
9. Contrast the differences between aperture cards and microfiche storage. When would you use each type of storage?
10. How does soft copy output differ from hard copy output?
11. Visit the data storage facility at your campus or a business firm, and describe the data storage devices in use.
12. Visit the data center at your campus, and describe the input media used. What type of verification, if any, is performed?
13. Call a local microfilm vendor in your area. Determine the cost for microfilming various types of documents.
14. Determine the approximate weight and volume of 10,000 hard copy docu-

ments, 8½″ x 11″. Compare the approximate weight and volume if this data were stored on microfilm.

15. If a microfilm reader is available at your campus, obtain a microfilm record and display it on the reader.

# Chapter 6

# Record, Forms, and File Design

Data processing systems can consume data in unbelievably large quantities. They can also produce an equally large, if not larger, amount of new data and information. Organizing and recording this data in a structured, manageable, efficient way is one of the most important tasks of the systems analyst.

Basically, the data related to a business system is organized in the following hierarchy:

| | |
|---|---|
| Item: | one piece of data or information |
| Record: | a group of related items |
| File: | a complete set of records |
| Library: | a collection of related files |
| Data bank: | the libraries used by a business firm |

## RECORD DESIGN AND LAYOUT

The basic unit the systems analyst works with is the record. It may be a source document containing information related to a transaction, an invoice showing details on a purchase, a punched card containing the hours worked by an employee, a portion of magnetic tape containing information relevant to an inventory item, or an area on a magnetic disk surface that records the current balance of a bank account.

Selecting and arranging the contents of a record is one of the responsibilities of the systems analyst. Several factors must be considered:

Available space on each record
Information to be included
Room to be allotted for each item
Order of the items

### Fixed and Variable Length Records

The length of a record determines how much data it can hold. Some records, such as punched cards or printed forms, have fixed lengths and can hold only a limited number of characters. A fixed length record consumes the same amount of space in a storage device regardless of how much data is recorded on it.

Other records have variable lengths, and consume only as much space as is needed to hold the required data. Some records may be short, and others long, depending on the length of the individual items recorded on each. Records on magnetic tape and disk media can be either fixed or variable in length. Variable length records are more efficient, but are often more complicated to use, since a means of identifying the end of each record must be developed.

### Design of Data Fields

The data to be included in a record is determined by the purpose of the record, the data that will be needed for processing, and how the information is to be obtained and accessed. An invoice record will contain information that is different from the information on a record showing a payment on a loan.

The source of data items to be recorded on a record must be considered, since they must be available in the right form and at the time the record is being prepared. Documents, other records, original transactions and activities are frequent sources.

The analyst must decide what size field should be allotted for each item. The size is a function of the information it will hold. For example, a field 20 characters wide may be reserved for a name. If a name requires only 12 characters, then 8 spaces will be left unused on that record. A name with more than 20 characters will be truncated. A field to record the year may be only two or four characters wide.

Field size is an important element when working with fixed length records or data that will be processed by some computer programs. It directly affects the selection of what and how much data can be included in a record.

Variable length records will, of course, be able to handle fields of differing lengths. A name with 12 characters will only require a field 12 characters wide. This increases efficiency, but a means of indicating the end of a field by a coded character or special word mark must be developed.

Another specification is the type of data that will be entered in a field. Some fields are restricted to numeric data, others to alphabetic data, and some to a combination—alphanumeric data. This depends on the format of the source data, the purpose of the record, and the way it will be processed and stored.

After defining the number and type of fields on a record, the analyst must select the basic sequence, or order, in which these fields will appear. This is largely influenced by how the record will be prepared, the order and availability of source data items, the order in which the fields will be accessed during processing, and the frequency with which each field will be accessed.

Fields on a record should be ordered to increase efficiency and accuracy as much as possible. Since most data on records is copied in one form or another from another record, it is more efficient for the data to be in the same sequence on both records. Fields that will be accessed frequently by machine or humans should be placed first on the records.

### Master and Detail Records

Records are often categorized into one of two groups, depending on the nature of the information they contain. Records with information of a relatively permanent or summary nature are called master records. Those with information relating to a specific transaction or activity are called detail or transaction records.

For example, a master record might include the name, address, phone number, social security number, exemptions, deductions, and pay rate of an employee, as well as other information of a relatively constant nature. The detail record would contain the number of hours worked, overtime, commission sales, business expenses incurred, or any other information related to a particular work week.

Records can also be defined by certain characteristics. A physical record refers to a specific, tangible record, such as a punched card or a particular portion of a magnetic tape or disk. A logical record refers to a related collection of items that may not be on the same physical record. It might refer to several adjacent physical records that contain information regarding the same subject.

## FORMS DESIGN AND LAYOUT

The systems analyst is often responsible for planning the different forms and records used by a business for recording source data, data input, processing, and data output.

## Business Forms

Many factors should be considered when designing business forms. These include type of data to be recorded, purpose of the form, method of data entry, and physical characteristics of the form itself. Forms should be easy to understand, convenient to use, and be designed to facilitate business activities and functions.

The most commonly used forms are:

1. CARBON INTERLEAVED FORMS. This is a multi-part form with a sheet of carbon paper inserted between each page. (See Figure 6.1.) With

**Figure 6.1.** Carbon interleaved form

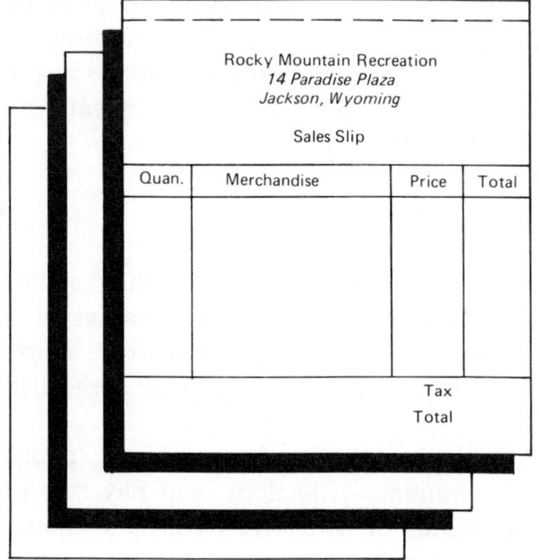


this form, duplicate copies are made as the original data is entered. Up to ten copies may be made; however, the quality of each succeeding copy degenerates somewhat.

Carbon interleaved forms are often supplied in a snap-apart style. In this format, all pages and carbons are fastened together at the top, creating a stub. When the form has been completed, the stub is pulled off, separating the carbon papers from the pages of the form.

This type of form is used for such things as paychecks with carbon copy vouchers or sales order forms with copies for different departments.

2. SPOT CARBON FORMS. This is a variation of the carbon interleaved form. Carbon is printed in selected areas on the back of each form, rather than being inserted between pages. (See Figure 6.2.) This allows only selected information to appear on the copies. Examples include invoices in which the shipper copy should not have the price, and paychecks with vouchers.

**Figure 6.2.** Spot carbon form

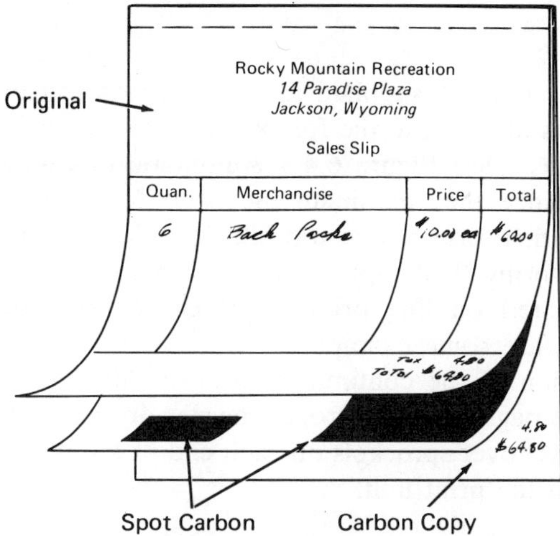


3. NCR PAPER FORMS. NCR (no carbon required) forms eliminate the need for interleaved carbon sheets or printed carbonized areas on multipart forms. A special chemical, which is a substitute for carbon, is applied to the face or back of each page. When the forms are written or typed upon, the impression appears on the duplicate copies. NCR forms are clean to use and eliminate the inconvenience of carbon papers. (See Figure 6.3.) They

**Figure 6.3.** NCR paper forms

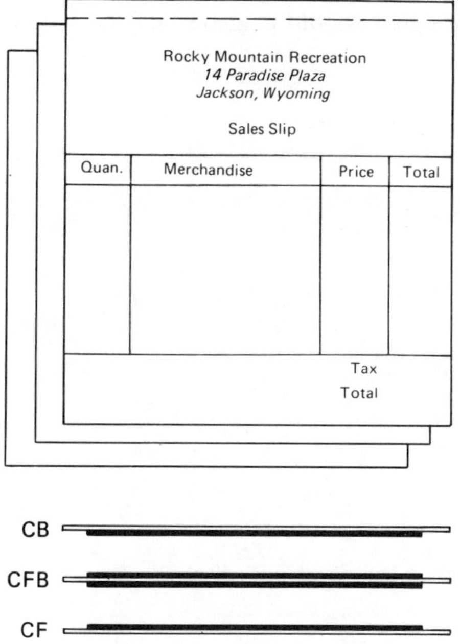

are often more expensive, however, due to the higher cost of the specially coated paper. This type is used for a whole range of business forms for which copies are needed.

CONTINUOUS FORMS. Forms may be supplied in a continuous roll or fan-folded in boxes. Each form is attached to the forms before and after it, separated by rows of perforations. (See Figure 6.4.) Automatic machines usually use continuous forms since they eliminate the need for manual insertion of each form. Once the first form has been positioned, the machine automatically feeds subsequent forms. This type is used for computer generated output. Paychecks prepared on line printers, order confirmation forms, invoices, and statements are some examples.

The pin feed form is a version of the continuous form. Small register holes punched at the sides of the page facilitate proper register during automatic feeding. The pin holes travel over sprockets on each end of the platen roll as the paper moves through the print unit.

PADDED FORMS. Forms may be ordered as pads, with or without chipboard backing. Pads of 50 or 100 forms are bound together with a rubber base adhesive at the top of the pages. Padded forms keep pages neat and in sequence. Numbered forms are often padded to facilitate using them in sequential or serial order. Padded forms are used for such things as interoffice memos, order form books, or job quotation forms.

### Graphic Considerations

Well-designed forms and records speed entering and processing of data and increase the level of accuracy. The designer should consider several factors when planning forms:

**Figure 6.4.** Continuous forms

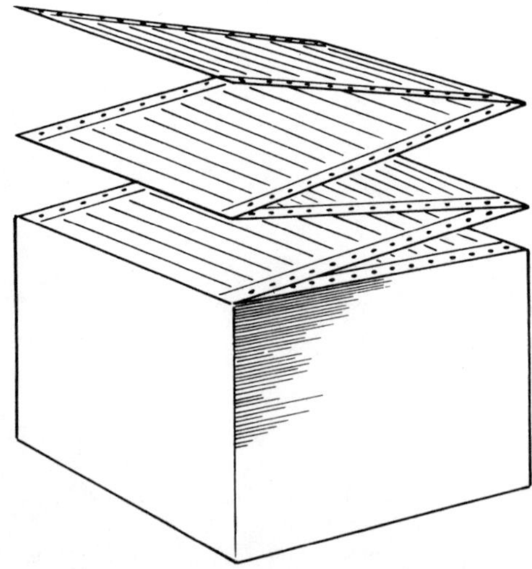

Paper selection. Forms can be printed on paper of many different colors, weights, stocks, and sizes.

1. Colored paper. Printing on colored paper is a convenient way of identifying copies. It facilitates sorting, or separating, parts of forms. A common color selection for forms is:

| | | |
|---|---|---|
| White | — | First copy |
| Yellow | — | Second copy |
| Pink | — | Third copy |
| Blue | — | Fourth copy |
| Buff | — | Fifth copy |
| Green | — | Sixth copy |
| Salmon | — | Seventh copy |

2. Weight of stock. Paper is available in different thicknesses, called weight or substance. The weight of the paper stock should be suitable to the form. One-time-only forms may be printed on lightweight stock to save storage room and paper costs. Forms that will be handled many times should be printed on durable, heavyweight stock. Most business forms are printed on 16- or 20-pound bond. Listed below are brief descriptions of the common weights:

| | | |
|---|---|---|
| 9-lb. bond | — | very thin, used for many-part forms, too thin for general use. |
| 11-lb. bond | — | thin stock, use for multi-part forms. |
| 13-lb. bond | — | thin stock, use for multi-part forms. |
| 16-lb. bond | — | lightweight stock for general forms use. |
| 20-lb. bond | — | medium weight stock for general forms use. |
| 24-lb. bond | — | heavyweight paper for durable forms. |
| 28–32-lb. ledger | — | heavier than 24-lb. bond, used for permanent ledger cards and records. |
| 90-lb. index bristol | — | lightweight card stock for durable records. |
| 110-lb. index bristol | — | medium weight card stock for permanent records. |
| 140-lb. index bristol | — | heavyweight card stock. |

3. Type of stock. Forms may be ordered in a variety of paper types. Newsprint stock is generally not suitable for forms work, but is sometimes used when cost is important and large quantities of forms are required. Generally, utility forms are printed on sulphite bond, a paper made entirely of wood pulp. Pen and ink forms and records of a permanent nature are usually printed on rag bonds. This paper has a content of 25 to 100 percent cotton fiber.

4. Size of form. The selection of form size depends upon several factors—use, processing, storage. As much as possible, form sizes should be standardized and kept to a uniform page size to facilitate handling, storage, and filing.

Before a large quantity of forms is ordered, the printer should be consulted regarding paper sizes. Often a fraction of an inch difference in size can mean a substantial saving in cost.

Some common form sizes and their use are:

- 3¼″ x 7⅜″ — standard 80-column punch card size, used for forms related to punched card processing systems.
- 5½″ x 8½″ — common statement size form, conveniently cuts from letterhead size sheet.
- 8½″ x 7″ — common invoice size.
- 7¼″ x 10½″ — military and government correspondence form standard (not widely used outside of these organizations).
- 8½″ x 11″ — standard letterhead size, best for most forms since standard paper handling devices or files are designed for this sheet.
- 8½″ x 14″ — standard legal size, slightly longer than regular letterhead size, and frequently used when form will not fit on 8½″ x 11″ sheet.

### Designing Forms

The placement of data is an important element in the development of efficient, serviceable forms. The design should be carefully planned for maximum ease of use and legibility. Figure 6.5 provides a checklist of items

**Figure 6.5.** Form check list

A quick and easy method of checking the efficiency and economy of any form—new or old—before placing your printing order. Read the text at the bottom of this sheet.

**NECESSITY** **OK ?**

1. Has the entire system been checked and would a written procedure for the use of this form help put it into more efficient operation? ____ ____
2. Are all copies of the form or report necessary? ____ ____
3. Have the actual users of this form been consulted for suggested improvements, additional requirements and possible eliminations? ____ ____
4. Can the data furnished by this form be combined with some other form or can some other form be eliminated or consolidated with it? ____ ____
5. Has everyone responsible for the form or the form system approved it? ____ ____

**PURPOSE** **OK ?**

6. If form is to be sent from one person to another, are proper spaces for "to" and "from" provided? ____ ____
7. Will routing or handling instructions printed on each copy be helpful? ____ ____
8. Should this form be consecutively numbered, or have a place for inserting a number? ____ ____
9. If this is an Outside Contact Form, should it be designed to mail in a window envelope? ____ ____
10. If this form is to take information from, or pass information to, another form, do both have the same sequence of items? ____ ____
11. Have we taken into consideration the number of forms which will be used in a given time (4 to 12 months)—the possibility of changes, and how long the form will remain in use? ____ ____

**SIZE AND ARRANGEMENT** **OK ?**

12. Is the size right for filing, attention value, ample room for information and to cut without waste? ____ ____
13. Is all recurring information

being printed, so that only variable items need be filled in? ___ ___

14. Has space been provided for a signature? ___ ___

15. Is spacing correct for handwriting or typewriting? ___ ___

16. Are the most important items, which should be seen first prominently placed? (Near the top, if practicable.) ___ ___

**WORDING** **OK ?**

17. Does the form, by title and arrangement, clearly indicate its purpose? ___ ___

18. Is there a proper space for the date? ___ ___

19. Is the form identified by company name and firm name or code number to aid reordering? ___ ___

20. If this is a revised form, can it be distinguished from the previous form? ___ ___

**PAPER AND PRINTING** **OK ?**
**(Specifications)**

21. Should the form be on colored paper to speed up writing, distribution, sorting and filing; to designate departments or branch offices; to indicate days, months or years; to distinguish manifold copies; to identify rush orders? ___ ___

22. Have we specified paper which will be thoroughly satisfactory, economical enough for form use, consistent in performance and surely available for later reorders? ___ ___

23. Is proper weight of paper used for original and each carbon copy? (Bond Substances 9, 13, 16 and 20. Ledger Substances 24, 28 and 32. Mimeo-Bond Substances 16 and 20. Spirit and Gelatin Duplicator Substances 16 and 20.) ___ ___

24. Are detailed specifications complete? (Paper, type, ink, rules, punch, perforate, score, fold, gather, pad, carbon sheet, stitch, etc.) ___ ___

25. Can other forms, printed on the same paper as this one, be ordered now to reduce production costs? ___ ___

26. Have requirements been estimated correctly and is the quantity to be ordered most economical? (Consider probability of revision and rate of use.) ___ ___

**Pt. #** **REMARKS on Points Questioned (?)**

***DATE*** ________________ ***19***____ ***SIGNED*** ________________________

**How to Use This Form**

Run through this list and appraise a new or revamped form point by point with an initial (rather than a check mark) either in the column headed "OK" or "?." This will help in working out the most efficient form use and specifications and the best working arrangement of items and copy.

Points marked (?) for further study can then be appraised systematically and discussed with those who will regularly use the form. Findings and further details can be elaborated upon in the column for "Remarks" at bottom of the second column.

To pin down responsibility, the person or persons giving the final OK should place their initials opposite the remarks. The whole Check List should be filed with a copy of the form for future reference.

pertinent to forms design. It questions such things as wording and paper and printing specifications. This list should be reviewed before preparing or ordering forms.

1. SPACING. Factors to be considered include spacing, arrangement of headings, size of type, and margins. If the form is to be used longhand, adequate space must be provided for fill-ins or completion items. If the form is to be used on a typewriter, it is easier to enter data if the spacing conforms to the standard six lines to the inch.

2. ORDER OF ITEMS. Important items should be placed at the top of the form. Fields on a record should be grouped so that related items are in adjacent positions. The order of fields should bear a logical relationship to the order in which the data is acquired or entered. Items copied from a source document or other form should be in the same sequence on both forms. Figure 6.6 illustrates a punched record and a source document from

**Figure 6.6.** Items punched into record from source document

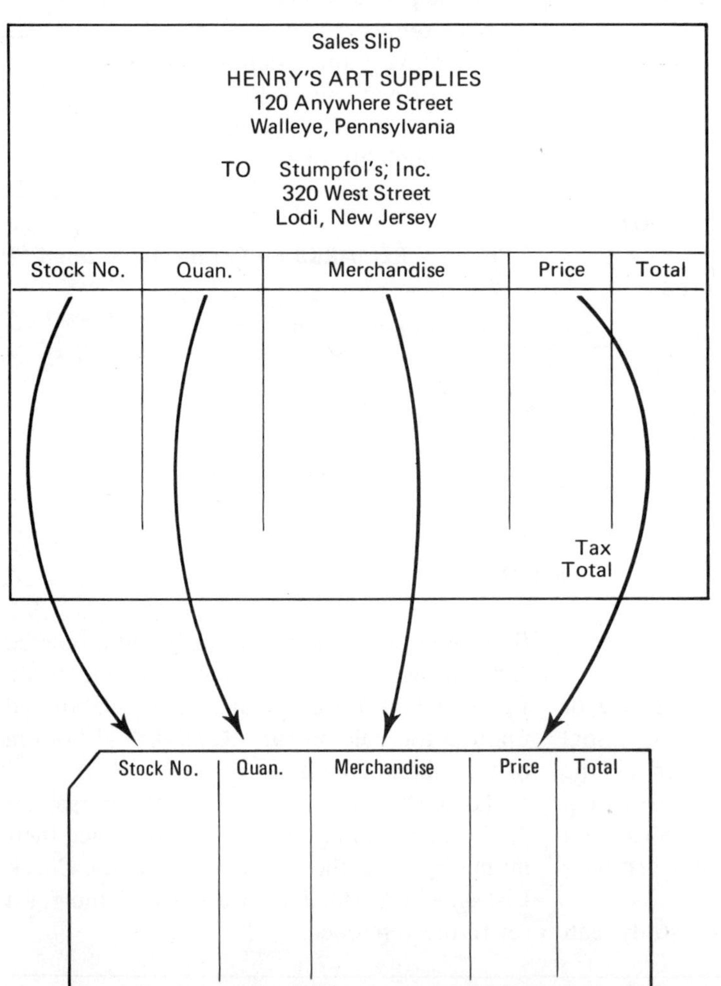

which the data was acquired. The flow of data entries on a form should generally be from left to right and from top to bottom.

A standard layout form is sometimes used in planning punched card records. (See Figure 6.7.) The form shows the layout to scale, including details such as spacing, card format, and interpreting. It serves both to document the card layout and as a record for the keypunch operator.

3. IDENTIFICATION. Output should be labelled. Documents such as reports, lists, and forms should be titled to identify purpose. All quantities shown in human readable form should be labelled to avoid confusion over what is being reported. This includes results of calculations and data retrieved from storage. Columns and rows should be clearly labelled whenever necessary.

4. MAXIMUM READABILITY. Reports, forms, and visual display of results should be designed for maximum readability. This requires attention to column spacing, placement of identifying text, or arrangement of output data. Lines should be widely spaced, and the columns well separated.

5. PHYSICAL CONSIDERATIONS. The graphic details of output reports and forms should be carefully planned. Adequate margins and space around text improves readability and facilitates entering data. Pages should always

**Figure 6.7.** Card layout form

IBM CARD LAYOUT FORM

FORM X74-4049-12

MATERIAL ACCOUNTING

MAT'L CLASS | STOCK NO. | DEPT. CHG. | ORDER OR ACCOUNT NO. | OPER. | QUANTITY REQ. | UNIT/M

DATE | COST PER UNIT | MAT'L CLASS | STOCK NO.

DEPT. CHG. | ORDER OR ACC'T. NO. | OPER. | QUANTITY REQ.

UNIT/M. | ITEM DESCRIPTION

FOREMAN'S SIGNATURE

MARK QTY. DEL'D. HERE

COMPLETE THESE SECTIONS FOR ALL LAYOUTS

COMPLETE THESE SECTIONS WHEN APPLICABLE

NEWARK — XYZ COMPANY — J. F. CRAWFORD, XYZ CO. — NEWARK, MONTANA — W. E. NORCROSS 4/4/6–

SCALE: APPROX. DOUBLE SIZE. ACTUAL CARD SIZE: 3 1/4" X 7 3/8"

Courtesy of IBM

be numbered, especially when multi-paged reports are being generated by a computer.

6. PERMANENCE. The output media should be sufficiently durable for its intended use. Forms and documents that are to be handled many times should be prepared on heavy paper. Information of a transitory nature may be output as soft copy on a video display screen.

7. LOGICAL SEQUENCE. Information being output in human readable form should be in the sequence and order used in normal business practices. For example, a measurement should be reported as 3′10″, not 10″3′. Some exceptions are occasionally necessary, such as the use of the Julian calendar instead of the conventional Gregorian calendar because of its adaptability to computer usage.

## CODING SYSTEMS

Coding systems may be used to save space and allow more information to be entered on a record or form. A code is assigned to represent data, and is entered on the record instead of the actual data. For example, two-digit numbers may be assigned to represent salespeople rather than spelling out their names. This requires only two characters on a record rather than 10 or 20.

Codes such as these should be sufficiently generalized to allow for expansion as new items are added, but specific enough to allow reference to each individual data item. The code should also bear a meaningful relationship to the item it represents. For example, a code to represent the months of the year should range from 1 to 12 rather than from 10 to 21. The coding system should be compatible with the type of processing—manual, machine, computer—that will be utilized.

The following systems are often used to code data on punched cards, as well as on other records:

1. SIGNIFICANT DIGIT CODE. In this system, each digit in the code represents a different characteristic of the data item. This arrangement allows considerable descriptive information to be recorded with a few characters. Example:

| Code | *Type of Media* | *Length in Minutes* | *Year Copyrighted* |
|---|---|---|---|
| TA 15 71 | Tape recording | 15 | 1971 |
| RE 08 70 | Disk recording | 08 | 1970 |
| FI 18 68 | Film, 16mm | 18 | 1968 |

2. SEQUENCE CODE. Another method of encoding data on a record is the sequence code. In this method, a group of items is assigned numbers in

sequence. Then the assigned number is used instead of the longer piece of data. Example:

| Code | *Item* |
|---|---|
| 01 | Carpets |
| 02 | Kitchen goods |
| 03 | Lighting fixtures |
| 04 | Hardware |
| 05 | Cosmetics |

3. MNEMONIC CODE. This system uses a shortened, or contracted, code symbol to stand for a longer term. The mnemonic code should bear a close relationship to the item it represents. For example:

| | |
|---|---|
| PAYPRO | Payroll Procedures |
| STINOP | Standard Inventory Ordering Procedures |
| EMTRED | Employee Training and Educational Department |

4. LAST DIGIT CODE. A final, or trailing, character is placed after a number or name to indicate the class to which it belongs. Either alphabetic letters or numeric characters may be used. Example:

| Code | *Meaning* |
|---|---|
| 103R | Retail item |
| 104W | Wholesale item |
| 105E | Export item |

5. IDENTIFIERS. Numbers, such as social security numbers and standard universal identifiers (SUI), are often used on records along with, or instead of, names. They are used as the identifying field to facilitate sequencing and searching records, and to avoid complications that arise from similar or duplicate names.

Since there are instances of duplicate social security numbers having been assigned, using this type of identifier is not without some risk. The analyst should also exercise care with social security numbers since their use is legally restricted to specific applications.

6. CHECK DIGIT CODES. Check digit codes are used to detect clerical errors that have been made when identifiers or other number codes are recorded. A check digit is a number added to the end of the code number. It represents the result of a mathematical formula that has been performed on the code number. The computer processing the data will perform the operations on the code number and compare the answer to the check digit. A disparity indicates that a transposition or other error has been made. Several formulas have been developed to calculate the check digit.

## FORMS CONTROL

The systems analyst is responsible for structuring procedures that facilitate forms control, which is the orderly, controlled movement of forms through a system. It is concerned with stocking, distributing, revising, printing, and specifying forms, input records, and source documents. Several factors should be considered when planning a forms control system:

1. Assignment of numbers. Each type of form or record should be numbered to ease reordering or referencing. A master list of forms used in any given system should be kept in a forms manual. The form number may include such information as the quantity printed and the date. Example:

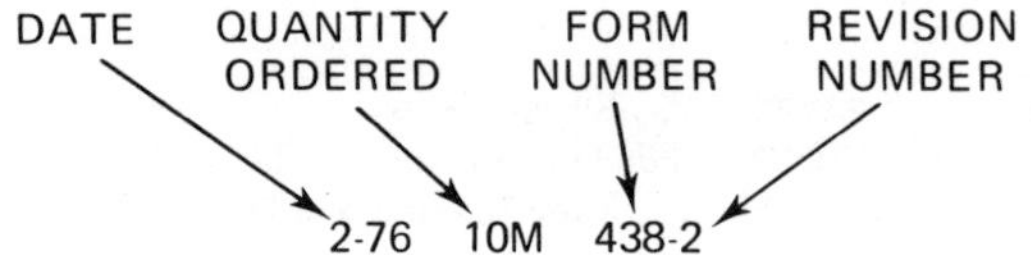


2. Revision number. If forms or records are revised or updated, a revision number should be assigned to the new form. This aids in keeping track of changes in content. All revisions should be recorded in the master forms book. One person should be assigned the responsibility of revising forms. All changes and modifications should be cleared through him or her to help avoid unnecessary increases in the number of forms and records in use.

3. Consecutively numbered forms. It is sometimes important to keep close control over each copy of a form. Consecutively numbered forms facilitate inventory control and help to maintain accuracy when handling returned merchandise transactions. A record should be kept of all numbers in use, and it should be checked before ordering new forms. Any new forms ordered from the printer should follow in the appropriate numbered sequence.

4. Routing of forms. If parts of a form are to be sent to different people or work stations, routing information should appear on each part. For example:

Original — Customer Copy
Duplicate — Accounting Department Copy
Triplicate — Sales Department Copy
Quadruplicate — Production Department Copy

Punched cards with different corner cuts or colored stripes can be used to identify records for various files or departments.

5. Instructions for use. Any form or record that is not self-explanatory or that requires an explanation for use should include descriptive text. Instructions may be printed directly on the form, or supplied separately with each packet. A copy of the instructions for completing a form should appear in the forms manual.

6. Inventory forms. The systems analyst will need to establish an inventory level to assure that an adequate supply of forms is always available. When a form reaches a predetermined minimum, it should be reordered promptly to avoid serious system problems. The forms manual should contain a record of the necessary information for reordering forms, such as supplier, cost, and specifications.

7. Deleting unnecessary forms. All forms and records should be reviewed periodically to assure that unnecessary ones are eliminated. The review should reexamine the need for a form, its routing locations, revisions, and improvements. Figure 6.8 shows a survey questionnaire that may be used for this purpose. It challenges each form on the basis of need. Forms that cannot be justified are systematically eliminated.

## FILE DESIGN

Records are grouped together in various relationships and orders to form files. The contents of a particular file depend on its function, the storage media utilized, and the processes it will undergo. Some files may be composed of master records, others of detail records, and still others of sets of records, each with one master record and several related detail records.

For example, a master file may contain all the records of customers with balances due. It might hold data such as name, address, charge number, and current balance. One detail file would include all charges made during the current month, and a second detail file would include all payments and returns made during that period. During processing, the information from these files would be merged and processed to produce a new master file, showing the new current balances.

Other examples would include an inventory file containing information related to each type of item in stock. The master record might show the information relating to supplier, wholesale and retail prices, reorder points, and stock number. The detail file might indicate the number of items sold by each order, or the number of items added to the inventory from a new shipment.

### Sequential and Random Order

The way in which records in a file are ordered depends on its purpose and the media involved. Records can be in sequential or random order. Sequential order means that the records follow each other successively, in a par-

**Figure 6.8.** Forms survey questionnaire

PAPERWORK SURVEY QUESTIONNAIRE
2 94293

DATE
FILE NO

1 FOR
☐ CONTRACT DATA REQUIREMENTS LIST (CDRL) ITEMS (DD 1423) ☐ NONCONTRACTUAL REQUIRED GOVT PAPERWORK
☐ INTERNAL (COMPANY) PAPERWORK ☐ PERIPHERAL GOVT DOCUMENTATION PAPERWORK

2 TITLE (AND NO.)

3 PURPOSE

| 4 AUTHORITY/REQUESTOR<br>NAVY ______<br>AIR FORCE ______<br>OTHER ______ | 5 COGNIZANT TECHNICAL OFFICE | | |
|---|---|---|---|
| 6 CLASSIFICATION (IF INTERNAL)<br>☐ INTRADEPART ☐ CORPORATE<br>☐ INTERDEPART ☐ OTHER | 7 MODEL APPLICABILITY | 8 ORIGINATING DEPT | 9. CONTRIBUTING DEPT |
| 10 ESTIMATED M/HRS TO ORIGINATE | 11 AVERAGE NO PAGES PER YR | 12 REPRODUCTION METHOD | 13 REPRODUCTION BY WHAT DEPT |
| 14 TOTAL QTY REPRODUCED<br>INTERNAL DISTR QTY ______<br>EXTERNAL DISTR QTY ______ | 15 DISTRIBUTION FREQUENCY | 16 DISTRIBUTION ESTABLISHED BY WHOM | 17 DISTRIBUTION CONTROLLED BY WHOM |
| 18 COST FOR CLERICAL, KEYPUNCH TERMINAL, ETC | 19 REPRODUCTION COST | 20 SPECIAL SUPPLIES COST | 21 DISTRIBUTION, FILING, DISPOSAL COST |

| 22 DATA PROCESSING COST | 23 COST PER<br>COPY ______<br>PAGE ______ | 24 TOTAL COST PER YEAR |
|---|---|---|

25 SOURCES OF INFORMATION-METHOD OF PREPARATION

26 OTHER PAPERWORK ITEMS PREPARED USING DATA FROM THIS ITEM

27 DOES THIS ITEM DUPLICATE OTHER ITEMS SUPPLIED INTERNALLY OR TO GOVT? ☐ YES ☐ NO
IF YES EXPLAIN

28 DOES COMPANY DATA EXIST WHICH WOULD SATISFY THIS GOVT REQUIREMENT? ☐ YES ☐ NO
IF YES EXPLAIN

PAPERWORK SURVEY QUESTIONNAIRE
2-94293 (BP)

TITLE (AND NO.) ______

DATE
FILE NO

29 IS VALUE OF THIS ITEM TO COMPANY OR GOVT QUESTIONABLE? ☐ YES ☐ NO
IF YES EXPLAIN

IF A REDUCTION OR ELIMINATION OF PAPERWORK ITEMS IS PLANNED OR HAS BEEN ACCOMPLISHED, AS A RESULT OF THE SURVEY, COMPLETE THE FOLLOWING:

30 CAN THE PAPERWORK ITEM BE ELIMINATED, SIMPLIFIED OR COMBINED. ☐ YES ☐ NO
IF YES LIST COST SAVINGS ______
(ATTACH COST SAVINGS WORKSHEETS)

31 CAN THE DISTRIBUTION OF ITEM BE REDUCED? ☐ YES ☐ NO
IF YES LIST COST SAVINGS ______
(ATTACH COST SAVINGS WORKSHEETS)

32 HAS THIS COST SAVINGS BEEN REPORTED? ☐ YES ☐ NO
IF YES TO WHOM

33 ADDITIONAL INFORMATION

| SIGNATURE OF PREPARER | DATE | APPROVAL SIGNATURE | DATE |
|---|---|---|---|

PAPERWORK DISTRIBUTION – QUESTIONNAIRE
(FOR EXISTING DOCUMENTS)
2 94289 R1

DATE

DOCUMENT NO ______

DOCUMENT TITLE ______

NO. OF PAGES ______ FREQUENCY ______

YOU ARE PRESENTLY ON DISTRIBUTION FOR THIS DOCUMENT AND CAN HELP **CONTROL** PAPERWORK **COSTS** BY OBJECTIVELY EVALUATING YOUR NEED FOR CONTINUED DISTRIBUTION

DO YOU FILE A COPY OF THIS DOCUMENT?

☐ YES ☐ NO

PLEASE INDICATE ACTION DESIRED
- ☐ No longer required Discontinue
- ☐ Continue Required for
  - ☐ Making decisions
  - ☐ Information only
  - ☐ Source material for other reports or documents
  - Other ______
- ☐ Change No. of Copies from ____ to ______
- ☐ Change frequency from ____ to ______

**FAILURE TO REPLY IN 10 DAYS WILL RESULT IN REMOVAL OF YOUR NAME FROM DISTRIBUTION**

(Use other side for additional comments)

______ NAME OF RECIPIENT ______ UNIT

______ SUPERVISOR'S SIGNATURE

Please sign and return this questionnaire to ______

FOR CONTRACTOR PERSONNEL USE ONLY

**Subject:** Paperwork Cost Review Monthly Report **Date:**

**To:** Executive Vice President **Memo:**

**From:**

The review of paperwork in

______
(Organization Name) (Organization Number)

has been effective in eliminating

______ Systems
______ Reports
______ Copies of Reports
______ Pages/sheets of paper
______ Forms
______ Other

This estimated annual savings amounts to: $ ______

In addition, simplification, improvement, and combining paperwork items has resulted in annual savings amounting to: $ ______

**Total estimated annual savings this month: $** ______

As a direct result of this reduction, we plan to implement the following course of action: ______

**Previously reported total annual savings: $** ______

**Accumulative annual savings to date: $** ______

Submitted by: ______
Title: ______
Ext. ______

ticular arrangement. This might be by the alphabetic, chronologic, or numeric order of one field on the record. Records being added to the file must be placed in the appropriate position in the file to maintain the sequential order.

Random order means that the records are placed one after the other with no regard to succession. Additional records are added to the file by placing them at the end or in the position vacated by a deleted record. Records in random order will need to have an identifying field containing key information, such as a number, name, or code.

Depending on the media used, records in a file can be accessed either sequentially or randomly regardless of how they are ordered. Any file can be searched sequentially until the appropriate information is found in the identification or key field on the record.

Media with random access permit a record to be located by going directly to the address where it is stored. The address will either be indicated by the key, or identification code itself, or be determined from a reference table.

Files that are being merged or processed simultaneously will have to be in the same, or corresponding, order. For example, suppose monthly statements were being prepared from a master file and two detail files. The records in all three files must be in the same order. Usually, sequence and matching tests are performed to assure that the appropriate detail and master records are merged.

### File Processing and Maintenance

File maintenance refers to the process of designing and creating a file and adding or deleting records to keep it current. Records in the file undergo several types of processing. They can be searched, merged, matched, altered, revised, deleted, and rewritten. Producing needed output may require that several files be processed and their information merged. Or the information on one file may be used several times to produce a variety of output in combination with several different files.

## LIBRARIES AND DATA BANKS

The last unit in the hierarchy of data organization is the data bank. A data bank is composed of a collection of libraries. A library is a collection of related files used by a firm. They might be all the files related to personnel, or all those used by the sales or inventory departments.

Data bank design involves storing data and records in a format that conveniently allows access to selected portions of the files. The particular structure of the data bank will depend on the kind of information to be stored and retrieved, the way the data is to be retrieved, the storage media used, the type of processing to be performed, and the type of output needed.

Data banks are often organized in categories to facilitate data retrieval. Each category has a reference name called a descriptor or keyword. Referencing the appropriate descriptor during data retrieval will access all the related data in that category.

## EXERCISES

1. List some of the factors that must be considered in designing business forms.
2. List several of the choices and factors involved in selecting the paper for forms.
3. What are the major factors to consider in record layout and design?
4. List several major coding systems and give an example of each.
5. What are the major considerations in planning output records?
6. What factors are normally considered in the forms control activity?
7. Gather several examples of business forms from firms with which you have contact. Write a brief evaluation of each, using the criterion discussed in this chapter.
8. Select one of the above forms and redesign it, improving graphics, layout, or content.
9. Obtain several samples of paper used in forms printing. Determine the type and weight. What recommendations for improvement would you make?
10. Visit a local printing firm and discuss the elements of forms design. What suggestions do they have to improve printed forms?
11. Select two related forms from exercise 7. Design a record to include the data from them.
12. Design three records that would be needed by the credit department of a firm that sells musical instruments and albums.
13. Specify the files that will be needed by the firm in exercise 12.
14. What type of media would you recommend for data storage for this firm? Why?
15. Select a business firm for study. Determine what files, records, and libraries should be used in building a data bank.

# Chapter 7

# Scientific Problem Solving

"Landing a spacecraft within 200 feet of its target on the surface of the moon is like throwing a needle across the Pacific Ocean into the eye of another needle located in Australia . . ." That is how an aerospace executive described a space age achievement made possible by a remarkably accurate guidance system and a most carefully conceived and executed data processing system.

The successful moon landing was a result of blending modern engineering and data processing expertise and the techniques of scientific problem solving—the laws of science, mathematics, and logic were used in a systematic way to solve a most difficult and incredibly complex problem.

The scientific method of problem solving is an important part of the systems analyst's repertoire, and is used to develop solutions for a wide range of business data flow problems. Its major advantages include:

1. REPRODUCIBILITY OF RESULTS. There is a good deal of assurance that procedures, operations, or tests carried out according to the scientific method will produce the same results each time they are performed.

2. ACCURACY OF RESULTS. Results based on the application of factual knowledge, reproducible experiences, and intellectual proficiency, rather than on guesswork or chance, will have a more reliable level of accuracy.

3. Efficient expenditure of time and effort. Since the procedures of the scientific method are executed systematically, there is more assurance that the activities of the systems analyst will remain relevant and directed toward specific goals.

4. Greater facility in handling complex problems. The systematic approach to problem solving gives the systems analyst a reliable and tangible plan to work from when handling a complicated problem with many diverse factors. It adds assurance that all relevant procedures, elements, and data will be considered, and that none have been accidentally omitted. It provides a precise plan to follow for coordinating and evaluating the factors involved in the study.

5. Transferability of results. Because results obtained from the scientific method of problem solving are reproducible, the systems analyst can expect similar results to occur in similar situations without having to repeat the steps in the process, or to retest data.

Scientific problem solving incorporates the following attributes and rules (which should be closely adhered to for best results):

1. Careful attention to details.
2. Objectivity in making observations.
3. Precision in analyzing data and reporting results.
4. Use of mathematics and statistical techniques.
5. Systematic logical plan for attacking problems.
6. Evaluation of results, used to readjust a system to bring it more in line with objectives.

## THE SCIENTIFIC METHOD

The basic scientific method of problem solving was described by philosopher John Dewey around the turn of the century when he outlined the steps of a logical approach for developing sound solutions to problems. These principles are employed whenever scientists, engineers, business people, designers, or others want to solve a problem with the least expenditure of time and effort, and achieve predictable, reproducible results. (The tools of scientific problem solving are discussed in Chapter 13.)

### Step 1. Recognize the Problem and Identify its Causes

A prerequisite for solving a problem is to define and delineate the elements involved in it. In this step, the system analyst describes and isolates the factors pertinent to a particular problem, and tries to determine the causes. The nature of the problem is investigated to see if it is a unique situation, indicative of a group of related problems, or if it is masking a more severe problem.

Suppose, for example, the agents in a large insurance company feel they are hampered in their work because the clerical department takes an excessive amount of time to prepare policies, letters, and other typed material. The first step will be to identify the factors involved and determine where the problem lies. Is it the fault of the clerical department? How does its output compare to that of similar departments? Are the agents at fault? Are they giving the clerical department sufficient data to prepare documents? Are the forms used to record data adequate and efficient? Is the problem a combination of two or more elements?

Let us assume that in this case the systems analyst decides that the level of output from the clerical department is below what reasonably could be expected. The next question is why? Is the low productivity due to inadequate equipment or malfunctioning typewriters? Is it inefficient office arrangements; is it problems in the physical environment such as high noise levels or lack of air conditioning? Is the difficulty due to typists with inadequate skills or insufficient training? Is it because the company employs too few typists to handle the load?

Once the real elements and causes of the problem have been identified, a solution can be planned. The problem cannot be solved if the wrong causes are attacked or the wrong solution is implemented. If the problem is due to inadequate typing skills, rearranging the office furniture will not increase the output.

### Step 2. Express the Problem in Precise Terms

The next step in the scientific method involves reducing the problem to quantitative terms. The systems analyst cannot be sure that a problem has been solved unless there is some objective way to measure results or the effects of changing variables within a system. In most instances, it is impossible to determine the best way to solve a problem until it has been clearly and specifically defined.

For example, the next step in this particular problem is to express both the present and expected levels of output in measurable quantitative terms. The systems analyst must know exactly how many pages below expectation the department falls, and the percentage of errors being made.

If the cause of the present low output is identified as due to typists with inadequate skills, then the next step is to determine exactly which skills are below the norm and by how much. To do this, tools that precisely measure typing performance must be obtained or developed. Various typing tests can be used to measure overall skills. Specially designed sample pages, policies, and forms and carefully planned dictation can be used to test the ability to perform specific tasks required by the company.

### Step 3. Analyze Choices and Select a Plan

In this phase, the systems analyst seeks to develop alternative solutions to the problem. Possible methods of improving the situation are studied and considered. In selecting a plan of action, an If-Then logic is used—*if* a given solution is implemented, *then* what will happen? One solution may greatly increase accuracy at a slight increase in cost. Another may increase output but not accuracy, at a substantial increase in cost. Another may increase output slightly and reduce cost at the same time.

After the possible alternative actions and their expected outcomes have been analyzed, the systems analyst selects the one that best meets the criterion. The choice will normally be the one that promises the greatest improvement at the lowest cost.

In this example, the analyst may decide that an in-service training program is the best way to upgrade the skills of the present staff. A recommendation may also be made to the personnel department that they hire only those individuals with higher levels of typing skills.

### Step 4. Implementation

After a solution has been selected, the next step is to carry it out. This phase may involve a considerable expenditure of time and money. Equipment vendors are called in, new employees hired, training courses developed, new forms printed, new equipment installed. Patterns of data flow may be changed and new procedures instituted, or old ones altered.

Implementing the solution in the sample problem will involve developing in-service instructional material and arranging for the typists in the clerical department to participate in the special training program. New standards and tests are drawn up to enable the personnel department to select individuals for employment who have a higher level of skills.

### Step 5. Evaluation

Probably the most important and often the most frequently overlooked phase of scientific problem solving is evaluating the results. In this step, the systems analyst determines whether the implemented solution actually solved the problem. The measures and criterion established in earlier steps are used to decide whether the expected improvements or changes actually took place.

This is an essential step in scientific problem solving. If outcomes are not tested and verified quantitatively, there is no tangible way to check the effectiveness of a solution. Without an accurate measure, it is difficult to to determine whether or not the goal has been reached, or whether more modifications in the solution are needed.

The systems analyst in the example re-administers the typing tests developed in Step 2 to evaluate the results of the training program and raised hiring standards. If the level of typing skills is up to expectations, then that part of the solution has been successful. If it is not, then more analysis and modifications are necessary.

The systems analyst will also reexamine the level of output produced by the clerical department to see whether or not it has improved. If it has not, then it is possible that the true cause of the problem has not been correctly diagnosed.

### Step 6. Feedback and Readjustment

In this step, the systems analyst reexamines the solution in light of the results of the evaluation. If the original expectations were not met, then a return to an earlier step in the problem solving procedure is indicated. The activities of that step will be repeated, usually with some modifications, in an attempt to adjust the outcomes. Then the analyst continues on with the succeeding steps.

If evaluation again shows that the results are not acceptable, the systems analyst returns once more to the appropriate earlier step. This process of reentering the problem-solving loop at the point indicated by the evaluation is called *feedback*. (See Figure 7.1.)

Few solutions produce perfect results the first time they are implemented or evaluated. This is especially true when the problem is relatively complex or when the solution involves making radical changes in a system.

The systems analyst will study the results of the evaluation and select the best point at which to reenter the problem-solving loop. Step 3 will be the reentrance point when modifications or changes are needed in the solution itself. If the problem lies in the way the solution was carried out, a return to Step 4 would be indicated. In some instances the analyst may decide that the evaluation itself was inaccurate and will repeat Step 5.

If these measures fail or the evaluation shows that the error is more fundamental, it may indicate a return to Steps 1 or 2 to completely reanalyze the problem. After reentering the loop at the appropriate point, the systems analyst repeats the processes to be performed in that step and then moves to the next sequential steps to complete the loop again.

Feedback from the repeated evaluation may show that more changes are needed, and the analyst will again reenter the loop at the appropriate point. This process continues until the results of the evaluation show that the goals have been met and that the problem has been solved.

## EARLY SCIENTIFIC MANAGEMENT

Frederick W. Taylor was one of the first to use the scientific method to solve problems in industry. He used its principles as a tool to learn the "one best way" to do a given task. He believed that greater output would result from a system if all employees followed the "one best way." Taylor,

Figure 7.1. Feedback loop

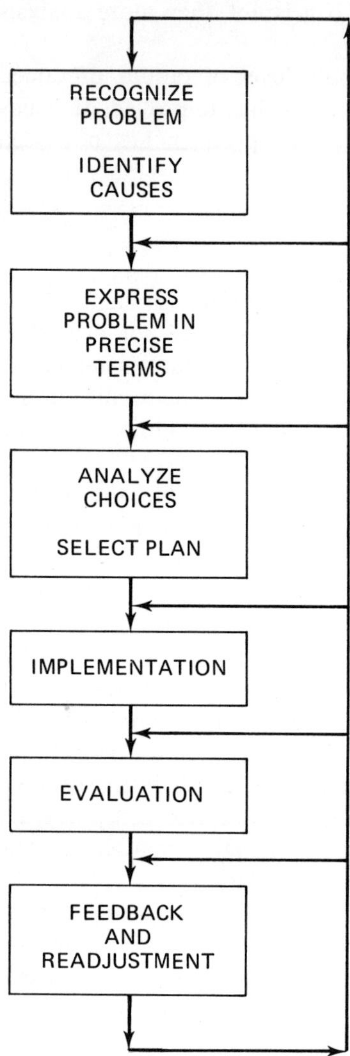


sometimes called the father of scientific management, felt it was management's responsibility to determine this "one best way," and to use rewards as incentives to train employees in its use.

This philosophy led, in the 1930's, to the stereotype of the "efficiency expert" with stopwatch and hidden camera . . . lurking in the shadows, spying on workers. Now we know that output depends upon many factors and influences that are not always as obvious and tangible as the "one best way" to perform a task. Psychological and attitudinal responses, peer group pressures, and status symbols, among others, are all important, related factors.

The accomplishments of the modern systems analyst has gone far beyond those of the efficiency expert. But the scientific method, with its sound, valid principles, remains a widely used, important systems analysis tool.

## SYSTEM STUDY AND INVESTIGATION

There are several methods and techniques that systems analysts use to gather the data necessary for analyzing, studying, and designing a system. These tools facilitate defining the elements of a system, the nature of a problem, and the knowledge and information necessary for its solution.

### Interviews

The face-to-face interview is a powerful tool of systems investigation for gathering information on activities, attitudes, and facts. Using structured and unstructured interview techniques, the investigator talks to employees, customers, or management, carefully recording the responses. Later, the gathered data will be analyzed, categorized, compared, interpreted, and so on.

The investigator often asks the six classical questions of the inquiring journalist during interviews—the 5 Ws and the H.

1. WHO? This question attempts to identify people, roles, and responsibilities. Who does this operation? Who is in charge? Who do you prefer to work with?

2. WHAT? This question is used to learn what elements or feelings are involved in a procedure or system. What are the activities in this department? What steps does the paperwork move through? What calculations are performed? What are the attitudes toward the tasks that must be done?

3. WHY? This question searches for the need or purpose of an operation and the reasons behind existing attitudes. Why is this job done in this way? Why is this operation done at all? Why do you feel that way?

4. WHEN? This question investigates time elements. How much time does it take to do a given job? When is it done? When do you feel tired?

5. WHERE? This question is concerned with physical location. Where is the station where this work takes place? Where is the storage area? Where is that activity done?

6. HOW? This question uncovers the sequence or order of steps followed in doing a job. How is this routine done? How is this activity performed? How did you learn to do this?

Interviews such as these yield a good deal of factual knowledge, as well as information on attitudes and feelings toward jobs, fellow workers, or superiors. The information enables the systems analyst to assess the behavioral responses of those involved in day-to-day working activities. Such attitudes are as important to the success of a system as are the physical manipulations and activities which the employees perform.

The factual knowledge gained is used by the investigator to build a picture of the interrelated elements of the system. It is difficult, or nearly impossible, to improve a system without a thorough understanding of its nature, strengths, weaknesses, and problems.

### Direct Observation

A second important means of gathering data is by direct observation. These techniques enable the systems analyst to gather additional knowledge about a system from a somewhat more objective base than can be found in personal interviews. Both viewpoints are, of course, necessary for a complete understanding of a system.

The systems analyst attempts to answer the same six questions described under interviews when making direct observations. Notes and records are carefully kept for later study and analysis. Analysts personally observe all aspects of a system. They watch people as they perform their tasks, taking careful note of such details as what they do, where they do it, how long it takes. They examine forms, documents, equipment, manuals. They watch the physical flow of movement as personnel walk from one work station to another when carrying out their duties.

The random clock observation technique is a common application of this method. At random times throughout the day (determined by a chart such as the one shown in Figure 7.2), the analyst observes and records the activities being carried out by personnel during that particular time. This technique allows the analyst to sample the activities of many days or weeks and condense their descriptions into a short, precise form suitable for further study.

Following is a typical breakdown of tasks developed during a random observation of a clerical worker:

| | |
|---|---|
| Talking to fellow employees | 15% |
| Telephone | 15% |
| Typing | 15% |
| Rest | 20% |
| Dictation | 10% |
| Talking to customers | 15% |
| Miscellaneous, unaccounted | 10% |
| Total | 100% |

### Review of Documentation

Another tool of the systems investigator is the methodical study and review of the official documents used in a system. The investigator obtains copies of policy and procedure manuals, personnel guides, bulletins, reports,

**Figure 7.2.** Random clock observation chart

| | | | | | | | |
|---|---|---|---|---|---|---|---|
| 10:29 | 8:00 | 9:11 | 11:55 | 4:04 | 1:19 | 10:35 | 8:05 |
| 1:10 | 8:16 | 4:22 | 4:29 | 2:17 | 8:25 | 3:53 | 11:38 |
| 2:07 | 10:40 | 3:48 | 10:10 | 11:46 | 2:15 | 3:25 | 11:49 |
| 3:01 | 10:46 | 9:08 | 4:42 | 2:09 | 3:45 | 2:51 | 11:57 |
| 11:36 | 8:30 | 9:35 | 4:20 | 11:51 | 4:06 | 1:40 | 11:50 |
| 9:50 | 2:47 | 3:59 | 2:24 | 11:31 | 8:10 | 11:42 | 3:36 |
| 10:17 | 10:12 | 8:15 | 8:38 | 9:42 | 10:19 | 3:04 | 1:33 |
| 2:03 | 10:41 | 10:53 | 1:38 | 2:19 | 9:06 | 11:34 | 11:13 |
| 4:37 | 1:28 | 11:11 | 2:41 | 9:38 | 1:27 | 4:56 | 8:47 |
| 2:20 | 2:25 | 10:52 | 1:02 | 9:12 | 9:46 | 4:01 | 1:56 |
| 9:56 | 9:51 | 10:23 | 9:23 | 8:54 | 3:22 | 4:23 | 1:51 |
| 4:55 | 2:38 | 11:32 | 2:46 | 8:45 | 4:21 | 1:29 | 4:19 |
| 1:20 | 1:04 | 9:39 | 9:45 | 10:02 | 3:14 | 9:07 | 11:10 |
| 2:02 | 8:50 | 10:43 | 10:18 | 8:12 | 9:37 | 3:30 | 10:58 |
| 1:34 | 2:27 | 3:49 | 10:09 | 10:34 | 11:54 | 1:01 | 9:17 |
| 2:54 | 10:27 | 8:20 | 3:31 | 9:32 | 3:44 | 2:18 | 3:43 |
| 2:10 | 11:53 | 11:09 | 4:28 | 4:09 | 4:25 | 4:24 | 8:03 |
| 9:16 | 11:16 | 2:06 | 4:18 | 8:33 | 4:12 | 2:26 | 11:39 |
| 4:47 | 11:37 | 2:08 | 9:54 | 3:52 | 9:31 | 11:05 | 10:22 |
| 9:10 | 8:35 | 8:06 | 2:28 | 8:36 | 2:33 | 3:24 | 10:28 |
| 4:38 | 11:58 | 8:19 | 11:59 | 2:12 | 2:52 | 8:44 | 2:55 |
| 9:04 | 9:20 | 9:58 | 1:26 | 4:48 | 3:17 | 10:03 | 2:34 |
| 8:41 | 10:38 | 2:00 | 3:57 | 1:16 | 10:42 | 11:26 | 3:09 |
| 3:00 | 11:06 | 11:56 | 10:11 | 1:37 | 1:18 | 2:23 | 8:49 |
| 9:40 | 9:05 | 4:16 | 3:54 | 8:58 | 1:46 | 8:11 | 1:48 |
| 4:13 | 2:53 | 3:20 | 8:22 | 11:00 | 4:30 | 2:16 | 3:05 |
| 8:17 | 2:40 | 8:42 | 4:43 | 3:40 | 4:46 | 1:42 | 8:31 |
| 3:39 | 10:00 | 10:47 | 9:36 | 11:27 | 11:52 | 3:56 | 3:03 |
| 1:39 | 3:10 | 8:46 | 1:58 | 4:49 | 9:59 | 2:29 | 1:54 |
| 9:28 | 4:36 | 9:30 | 10:32 | 4:40 | 10:33 | 11:02 | 8:40 |
| 9:43 | 2:42 | 9:53 | 2:36 | 3:38 | 10:31 | 8:59 | 3:33 |
| 3.12 | 1:47 | 2:37 | 4:51 | 11:29 | 9:00 | 8:14 | 11:24 |
| 8:07 | 4:27 | 9:47 | 2:50 | 9:15 | 8:37 | 10:20 | 9:26 |
| 8:34 | 1:13 | 3:27 | 1:17 | 3:26 | 9:13 | 11:01 | 3:06 |
| 1:00 | 1:23 | 2:58 | 4:00 | 4:34 | 1:52 | 3:32 | 3:50 |
| 10:07 | 4:52 | 11:25 | 3:58 | 10:36 | 9:14 | 10:16 | 11:19 |
| 11:30 | 10:24 | 11:45 | 10:30 | 1:39 | 4:03 | 2:30 | 1:21 |
| 3:19 | 9:18 | 4:32 | 11:44 | 10:37 | 1:55 | 8:23 | 8:01 |
| 11:03 | 3:08 | 3:34 | 8:48 | 3:16 | 2:45 | 3:41 | 11:41 |
| 1:05 | 10:45 | 1:57 | 10:08 | 4:45 | 3:55 | 10:48 | 2:21 |
| 9:19 | 9:33 | 10:15 | 1:11 | 10:57 | 10:25 | 9:03 | 10:54 |
| 11:28 | 8:51 | 3:18 | 1:14 | 3:23 | 3:42 | 9:22 | 1:41 |
| 8:02 | 11:33 | 4:58 | 3:02 | 11:47 | 8:53 | 4:08 | 9:09 |
| 9:01 | 4:05 | 10:59 | 8:32 | 11:07 | 2:39 | 1:32 | 10:05 |
| 9:41 | 11:21 | 2:56 | 2:48 | 4:50 | 4:31 | 4:14 | 1:22 |
| 9:48 | 8:29 | 3:21 | 8:28 | 3:46 | 11:14 | 1:30 | 9:55 |
| 1:53 | 4:33 | 1:07 | 10:56 | 8:43 | 8:24 | 1:50 | 3:29 |
| 2:49 | 1:24 | 9:02 | 3:47 | 11:35 | 10:39 | 9:29 | 9:57 |
| 3:35 | 1:25 | 2:32 | 8:39 | 3:51 | 11:43 | 3:07 | 3:11 |
| 4:27 | 10:44 | 4:53 | 10:13 | 8:18 | 4:35 | 8:08 | 4:41 |
| 1:45 | 11:18 | 8:57 | 2:04 | 1:36 | 10:49 | 4:39 | 11:48 |
| 8:09 | 4:15 | 8:56 | 9:24 | 1:44 | 9:44 | 4:07 | 11:23 |
| 1:03 | 4:59 | 1:59 | 2:22 | 1:06 | 1:31 | 10:55 | 2:31 |
| 10:14 | 4:57 | 2:14 | 2:35 | 8:27 | 3:13 | 10:50 | 4:26 |
| 1:09 | 4:54 | 10:01 | 8:52 | 10:26 | 1:15 | 8:04 | 11:20 |
| 2:57 | 8:55 | 1:43 | 4:10 | 10:51 | 9:34 | 1:49 | 11:08 |
| 2:05 | 11:12 | 4:44 | 8:13 | 2:01 | 4:11 | 10:21 | 4:17 |
| 3:28 | 1:08 | 10:06 | 2:11 | 2:43 | 2:44 | 11:15 | 2:59 |
| 8:25 | 3:37 | 11:04 | 1:12 | 11:40 | 11:17 | 10:04 | 8:21 |
| 9:25 | 2:13 | 11:22 | 9:49 | 3:15 | 4:02 | 9:52 | 9:21 |

Source: *The Office*, April 1972, p. 62

official forms, and other similar documents, and examines their functions and purposes. A review of this type gives an indication of whether the operations of a system reflect the official goals, and it helps to pinpoint existing omissions, weaknesses, or errors.

## Audits

A review of records, ledgers, files, memos, and other pieces of day-to-day paperwork also uncovers a considerable amount of information on the elements of a system and how they interrelate. This is often done by requesting personnel to save or prepare a copy of all working documents, notes, or records generated over a given period—say several days or weeks. Special provisions must often be made to facilitate gathering information from transitory documents such as memos.

The information from these papers is analyzed and used in several different ways. For example, data flow patterns in a system can be studied in this manner. The efficiency, value, or weaknesses of forms can be examined and rated. Data gathered in this way can be used to develop a work distribution pattern, called an action paper study. This technique plots the kinds of clerical tasks that are performed and shows the percentage of time involved. A typical breakdown for a clerical employee for one day might look like this:

| | |
|---|---|
| Typing, finished work reports | 30% |
| Typing, draft reports | 15% |
| Correcting, altering drafts | 20% |
| Typing letters | 15% |
| Typing memos | 15% |
| Typing stencils, masters | 5% |
| Total | 100% |

Information of this type serves as an excellent guide for the systems designer when selecting typing equipment for that particular work station.

## Questionnaires

Another important means of gathering information regarding a system is the use of questionnaires. Employees, customers, management and others may be asked to complete questionnaires designed to focus on general or specific problems in a system.

Short answer, true or false, fill-in forms, or multiple choice questions are valuable for eliciting specific data on a particular problem. (See Figure 7.3.) Open-ended (essay type) questions are often used to gather information of a more general nature or when an assessment of attitudes and interest is being made.

## Figure 7.3. Questionnaires

**PAPERWORK SURVEY QUESTIONNAIRE**

2-94291 (BP)

TITLE (AND NO.) ______________________ | DATE | FILE NO

29. IS VALUE OF THIS ITEM TO COMPANY OR GOVT QUESTIONABLE? ☐ YES ☐ NO
IF YES, EXPLAIN

**IF A REDUCTION OR ELIMINATION OF PAPERWORK ITEMS IS PLANNED OR HAS BEEN ACCOMPLISHED, AS A RESULT OF THE SURVEY, COMPLETE THE FOLLOWING:**

30. CAN THE PAPERWORK ITEM BE ELIMINATED, SIMPLIFIED OR COMBINED. ☐ YES ☐ NO
IF YES, LIST COST SAVINGS ______________________
(ATTACH COST SAVINGS WORKSHEETS)

31. CAN THE DISTRIBUTION OF ITEM BE REDUCED? ☐ YES ☐ NO
IF YES, LIST COST SAVINGS ______________________
(ATTACH COST SAVINGS WORKSHEETS)

32. HAS THIS COST SAVINGS BEEN REPORTED? ☐ YES ☐ NO
IF YES, TO WHOM

33. ADDITIONAL INFORMATION

| SIGNATURE OF PREPARER | DATE | APPROVAL SIGNATURE | DATE |
|---|---|---|---|
| | | | |

**PAPERWORK DISTRIBUTION – QUESTIONNAIRE**

**(FOR EXISTING DOCUMENTS)**

2 94289 R1

DATE

DOCUMENT NO ______________________

DOCUMENT TITLE ______________________

______________________

NO. OF PAGES __________ FREQUENCY __________

YOU ARE PRESENTLY ON DISTRIBUTION FOR THIS DOCUMENT AND CAN HELP **CONTROL** PAPERWORK **COSTS** BY OBJECTIVELY EVALUATING YOUR NEED FOR CONTINUED DISTRIBUTION

DO YOU FILE A COPY OF THIS DOCUMENT?

☐ YES ☐ NO

PLEASE INDICATE ACTION DESIRED

- ☐ No longer required  Discontinue
- ☐ Continue  Required for
  - ☐ Making decisions
  - ☐ Information only
  - ☐ Source material for other reports or documents
  - Other ______________________
- ☐ Change No. of Copies from _____ to __________
- ☐ Change frequency from ____ to ______________

**FAILURE TO REPLY IN 10 DAYS WILL RESULT IN REMOVAL OF YOUR NAME FROM DISTRIBUTION**

(Use other side for additional comments)

______________________ ______________
NAME OF RECIPIENT UNIT

______________________
SUPERVISOR'S SIGNATURE

Please sign and return this questionnaire to

______________________

**FOR CONTRACTOR PERSONNEL USE ONLY**

**PAPERWORK SURVEY QUESTIONNAIRE**

2 94293

DATE | FILE NO

1. FOR
☐ CONTRACT DATA REQUIREMENTS LIST (CDRL) ITEMS (DD 1423) ☐ NONCONTRACTUAL REQUIRED GOVT PAPERWORK
☐ INTERNAL (COMPANY) PAPERWORK ☐ PERIPHERAL GOVT DOCUMENTATION PAPERWORK

2. TITLE (AND NO.)

3. PURPOSE

| 4. AUTHORITY/REQUESTOR<br>NAVY ________<br>AIR FORCE ________<br>OTHER ________ | 5. COGNIZANT TECHNICAL OFFICE | | |
|---|---|---|---|
| 6. CLASSIFICATION (IF INTERNAL)<br>☐ INTRADEPART ☐ CORPORATE<br>☐ INTERDEPART ☐ OTHER | 7. MODEL APPLICABILITY | 8. ORIGINATING DEPT | 9. CONTRIBUTING DEPT |
| 10. ESTIMATED M/HRS TO ORIGINATE | 11. AVERAGE NO. PAGES PER YR | 12. REPRODUCTION METHOD | 13. REPRODUCTION BY WHAT DEPT |
| 14. TOTAL QTY REPRODUCED<br>INTERNAL DISTR. QTY ______<br>EXTERNAL DISTR. QTY ______ | 15. DISTRIBUTION FREQUENCY | 16. DISTRIBUTION ESTABLISHED BY WHOM | 17. DISTRIBUTION CONTROLLED BY WHOM |
| 18. COST FOR CLERICAL, KEYPUNCH, TERMINAL, ETC | 19. REPRODUCTION COST | 20. SPECIAL SUPPLIES COST | 21. DISTRIBUTION, FILING, DISPOSAL COST |

| 22. DATA PROCESSING COST | 23. COST PER<br>COPY ________<br>PAGE ________ | 24. TOTAL COST PER YEAR |
|---|---|---|
| | | |

25. SOURCES OF INFORMATION/METHOD OF PREPARATION

26. OTHER PAPERWORK ITEMS PREPARED USING DATA FROM THIS ITEM

27. DOES THIS ITEM DUPLICATE OTHER ITEMS SUPPLIED INTERNALLY OR TO GOVT? ☐ YES ☐ NO
IF YES, EXPLAIN

28. DOES COMPANY DATA EXIST WHICH WOULD SATISFY THIS GOVT REQUIREMENT? ☐ YES ☐ NO
IF YES, EXPLAIN

**Subject:** Paperwork Cost Review Monthly Report **Date:**

**To:** Executive Vice President **Memo:**

**From:**

The review of paperwork in

______________________________________
(Organization Name) (Organization Number)

has been effective in eliminating

__________ Systems
__________ Reports
__________ Copies of Reports
__________ Pages/sheets of paper
__________ Forms
__________ Other

This estimated annual savings amounts to: $ ________

In addition, simplification, improvement, and combining paperwork items has resulted in annual savings amounting to: $ ________

**Total estimated annual savings this month: $** ________

As a direct result of this reduction, we plan to implement the following course of action: ______________________

______________________________________

______________________________________

______________________________________

______________________________________

______________________________________

**Previously reported total annual savings: $** ________

**Accumulative annual savings to date: $** ________

Submitted by: ______________________
Title: ______________________
Ext. ______________________

## Time and Motion Studies

Another method of fact finding, which dates back to the early efforts of efficiency experts, is the time and motion study. This technique attempts to document and measure accurately the clock times and physical motions involved in performing a given task. Stopwatches and cameras are frequently used. Slow motion and single frame projections of the recorded motions are invaluable for studying and analyzing the steps involved in the various operations.

For example, a department store wanted to improve the procedures related to the cash register—handling purchases, returned merchandise, and checking credit over the telephone. A systems analyst had a 16mm camera with a telephoto lens installed and focused on a selected cash register. The camera was located in an unobtrusive spot, many feet away from the register. It was equipped with a mechanism that projected the image of a stopwatch on each frame of the motion picture, photographing the actions along with the times required for their performance.

The films were replayed later for careful study and review. The data learned from these films gave clues for needed changes in procedures and physical arrangements. After the redesigned system was implemented, the same camera equipment was used to document the new procedures and determine the actual time savings that resulted.

## Cost Analysis

A fundamental tool used in systems analysis is the cost study—an examination of the elements of a system based on the cost of processing a given unit of work. Discrete tasks, such as typing a letter, processing a phone inquiry, or replacing a file in a cabinet, were analyzed and measured to establish quantitative standards for comparison and study. This technique involves breaking down a procedure into its fundamental operations and determining the costs for each operation.

Later, the costs for various procedures can be calculated by adding together the costs of the individual operations involved. Predictions of costs can be made from these figures and then compared to the actual costs encountered. Costs of new and old systems can be compared and evaluated.

## Statistical Analysis

Other types of information describing a system can be gathered using modern statistical methods, mathematical techniques, and the computer. This type of quantitative data is an excellent means of monitoring the operation and functioning of a system.

For example, telephone companies often keep careful records on the speed of their service. They sample such things as dial tone delay, dialing times, connect time, and compute the average and standard deviation for a given period. Comparing these rates with those of other periods enables them to monitor the performance of the system over various intervals of time. Figure 7.4 shows the mean (average) connect time and standard deviation for calls of varying distances for a given period of time. These statistics are from a 1966 study and may not reflect current averages.

**Figure 7.4.** Statistical measures of Telephone calls

Connect Times for DDD Calls

| *Airline Distance (Miles)* | *Connect Time (seconds)* *Mean* | *Std. Dev.* |
|---|---|---|
| 0–180 | 11.1 | 4.6 |
| 180–725 | 15.6 | 5.0 |
| 725–2900 | 17.6 | 6.6 |

Source: Publication No.: PUB41005, American Telephone and Telegraph Co., Engineering Director —Transmission Services, 195 Broadway, New York, N. Y. 10007, page 12. These figures are currently being updated.

Other statistical techniques, such as correlations, chi square, and regression analysis, are used to uncover relationships not readily apparent from a cursory examination of the data.

### Communications Traffic Survey

A traffic study is a technique used to analyze the communications links that tie together many parts of a system. Businesses rely heavily on telephone and data transmission lines to join offices, branches, and departments.

However, this communications link frequently becomes a bottleneck, limiting the flow of information between units of an organization. The traffic survey is a valuable tool in such situations to pinpoint problems. The number and types of messages transmitted, their destinations, duration, or line charges for a period of time are carefully recorded. This information is invaluable in producing an overall picture of the communications flow in a given system that can be studied, evaluated, and manipulated.

## EFFECTS ON ENTIRE ORGANIZATION

Another important aspect of systems investigation is the study of the interrelationships of the parts of the system. The analyst must predict the effects that changing one part of a system will have on other parts. How will new equipment affect cost factors? What will be the change in morale if the structure of the business organization is reorganized? What other parts of the system will be affected if inventory and credit department data is computerized? The human element is always a major part of a system that must be considered whenever changes or alterations take place. But since the human element does not always respond predictably, it is often the most difficult one to plan for and evaluate.

### The Hawthorne Effect

The results of a study carried out during the 1920's well illustrates the complexity of human reactions. Psychologist Elton Mayo was studying a group of relay assemblers in the Hawthorne, Chicago, plant of the Western Electric Company, with the goal of increasing productivity. He performed a series of tests in which elements in the working environment were changed—lighting levels were manipulated, workers were isolated, rest periods were altered, and so on.

The final evaluation showed quite unexpected test results. Mayo had anticipated that productivity would go up as environmental conditions improved, and would go down as they declined. What actually happened was that the productivity of the study group went up regardless of what changes were made. Mayo concluded that people involved as subjects in a study often behave differently than they do when not under study. The very fact that someone knows he or she is involved in a study and is being monitored will affect his or her actions and output. This phenomenon is known as the Hawthorne Effect.

The astute systems analyst is aware of the psychological implications of this reaction when considering human elements in system change and design.

## DETERMINING CONSTRAINTS AND PARAMETERS OF SYSTEMS

Few systems operate without limitations or constraints, and a major portion of the analyst's time is spent in defining and evaluating where they lie. Business resources are limited, and the analyst must know how much money, time, personnel, equipment, or facilities should be applied to solve a particular problem. The rate of tolerable errors and length of time delays must also be determined.

The demands placed upon a system and the dependence and reliance upon it by others must be considered before attempting any changes or modifications. For example, the power company delivering electricity to a major city cannot arbitrarily shut off its power plants and generation equipment for several weeks to change or update its equipment. Other arrangements must be made which take these responsibilities into consideration.

Similarly, a business system cannot be closed or put out of service or disorganized for any length of time without some economic consequences. These must be predicted and planned for by the systems analyst when altering a system or installing a new one. It is often necessary to develop backup or alternative systems for use during the time that the original system is being modified.

Finally, a system must have facilities for incorporating future plans and handling projected growth. A system designed to satisfy only present needs will not suffice for long. Eventually, another, more adequate, system will have to be developed and implemented—increasing the costs related to systems analysis and greatly decreasing its effectiveness.

## EXERCISES

1. List four advantages of using the scientific method to solve problems.
2. Describe the characteristics or attributes of the scientific method.
3. What activities does the systems analyst carry out in the first phase of scientific problem solving?
4. What activities are carried out during the implementation phase of problem solving?
5. Why is the evaluation phase important in the problem-solving activity?
6. Describe the role Frederick W. Taylor played in early scientific management.
7. List several tools used in defining a system and the nature of the problem.
8. What kinds of questions are asked in the interview stage?
9. Describe how the random clock observation technique is used in system study.
10. How are time and motion studies used in analyzing a system?
11. What is the Hawthorne Effect and how does it affect system results?
12. Why should system parameters and constraints be studied before planning a new system?
13. Select a problem such as improving your reading or typing skills. Relate the solution to this problem to the scientific method.
14. Visit a business firm and determine which scientific problem-solving tools are used in solving their problems.
15. Select a simple clerical operation that may be observed in a business office. Perform an elementary time and motion study on this activity.

# Chapter 8

# Elements of System Flowcharting

A system flowchart is a graphic illustration of the data flow, or movement of information, through a system. System flowcharts are widely used by system analysts for planning, describing, troubleshooting, and documenting a system. The importance of such charts increases as systems expand in size and complexity.

This chapter covers the fundamentals of system flowcharting, illustrates and describes the symbols used in their preparation, and discusses several flowcharting examples.

## FLOWCHART DEFINITION

A flowchart is a graphical representation of the definition, analysis, or solution of a problem, in which symbols are used to represent such things as operations, data, flow, and equipment. System flowcharts are diagrams or visual graphic pictures that trace the movement of data from origination to end product. They show the sequence of steps through which the data moves as it is manipulated. A chart describes, in bird's eye form, the elements in a system, including related personnel, work stations, forms and records, processing, and associated activities.

### Program Flowchart

The systems analyst often uses another type of flowchart, called a program flowchart, when working with computer programs that perform specific functions. Program flowcharts document computer programs. They show the specific steps performed by the instructions in a program as they are executed.

System flowcharts document an entire system, of which the computer program may be one small part.

## SYMBOL DESCRIPTION AND USAGE

Flowcharts use a standard set of symbols, shown in Figure 8.1. These symbols have been approved by the American National Standards Institute, Inc. (ANSI), X3.5-1970. Both the shape of the symbol, and the text within it, convey descriptive information about the step being represented.

ANSI separates the symbols into several categories. The two major groups are basic and specialized symbols. Acceptable flowcharts may be drawn using only the basic symbols. Specialized symbols are more specific and give more information about an operation. They are, therefore, preferable to basic symbols.

### 1. Basic Input/Output Symbol

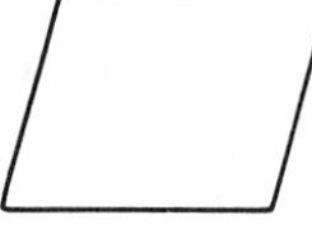

A parallelogram is the symbol that represents input or output media or operations. This is a generalized form and may be used for a variety of input and output media: line printers, card readers, card punches. It also indicates all types of input or output processes: reading in a payroll from time cards, reading sales slips, typing out invoices or reports.

### 2. Specialized Input/Output Symbols

Each specialized symbol represents a specific device used for an input/output function.

ONLINE STORAGE. The online storage symbol refers to input/output operations performed on electronic media that are connected to a computer and capable of storing or retaining the data. The symbol is used to indicate that a file is recorded on an online storage device, that new data is to be written on the device, or that data already recorded is being accessed.

This symbol may represent magnetic tape, disk, drum, or data cell units. Examples include preparing the monthly billing from files, and programs recorded on an online storage device.

**Figure 8.1.** Summary of flowchart symbols

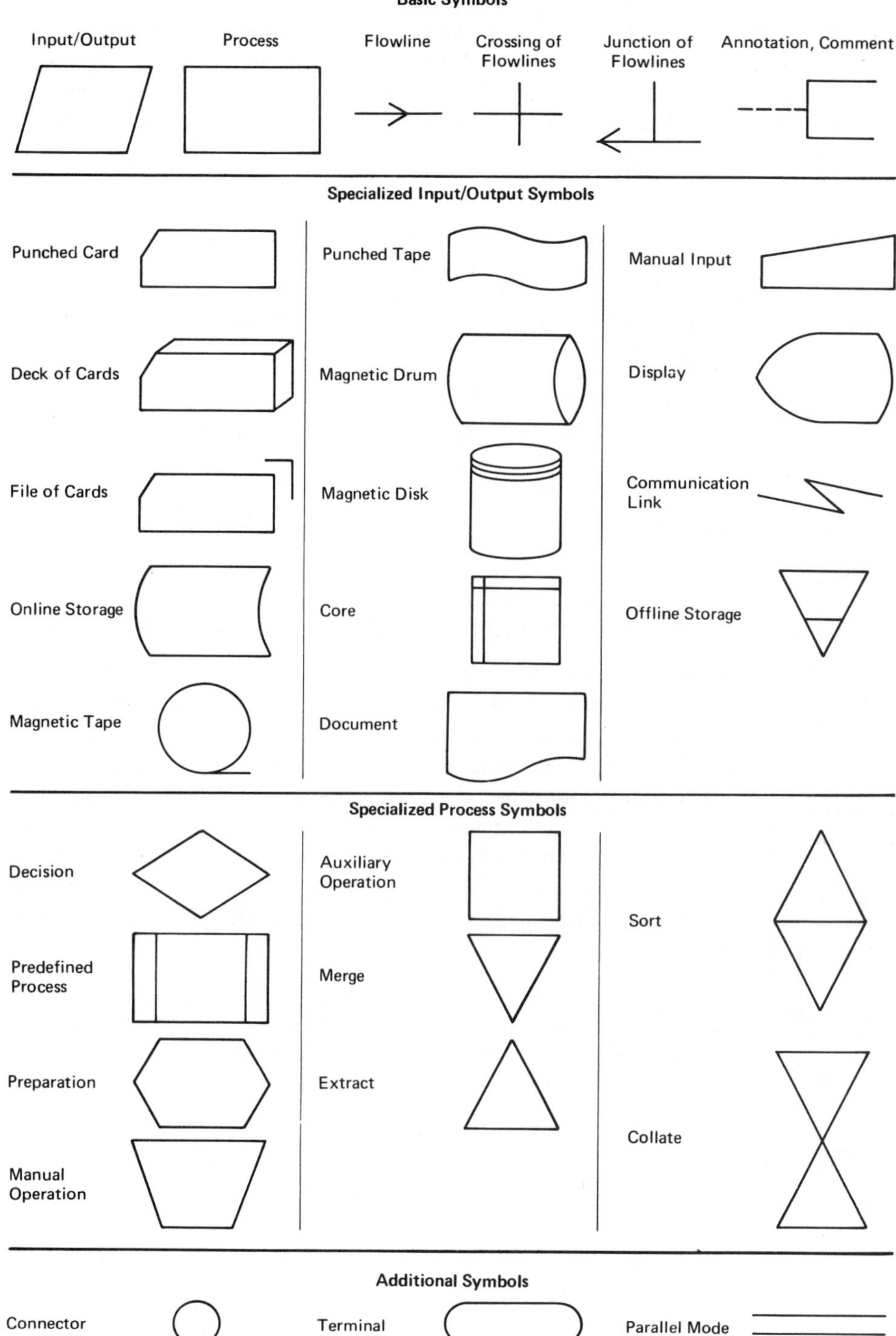

The following four symbols may also be used to indicate the particular online device:

MAGNETIC TAPE. A symbol shaped like a reel of tape specifies an input or output operation performed on magnetic tape. Examples include a parts inventory list stored on magnetic tape and a file of students' names on magnetic tape being used to generate a report. Magnetic tape can also be used offline with magnetic tape readers.

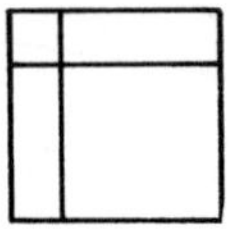

MAGNETIC CORE. This symbol specifies that information is being input or output from a magnetic core, or primary storage, device. Examples include a list of parts and their wholesale costs stored in the core memory system of a computer, or a quality control program stored in a minicomputer.

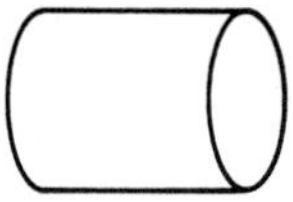

MAGNETIC DRUM. A cylinder-shaped symbol specifies input or output operations being performed on a magnetic drum storage device. An example would be a computer program that prepares a payroll stored on a magnetic drum unit.

MAGNETIC DISK. A symbol resembling a group of disks specifies input or output operations on magnetic disk storage devices. Examples include files such as alphabetic lists of employees' names, or professional periodicals recorded on a magnetic disk medium.

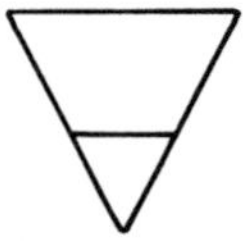

OFFLINE STORAGE. This symbol refers to input/output operations performed on data storage media not connected to a computer. It is used to indicate that a file is stored on that media, that new data is being recorded, or that stored data is being accessed.

The symbol may be used when the storage media is punched cards or tape, documents, or the like. Examples include transferring the data from employee time cards onto punched cards, or to indicate that customers' signatures are recorded on documents.

More specific symbols may also be used to indicate offline storage media.

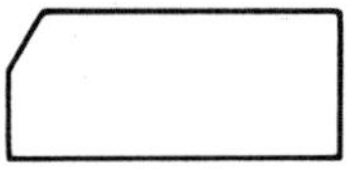

Card

Deck of cards

File of cards

PUNCHED CARDS. A symbol shaped like a punched card is used to show that data is stored or being input or output on punched cards. There are several forms of this symbol to indicate a single card, a deck of cards, or a file of cards.

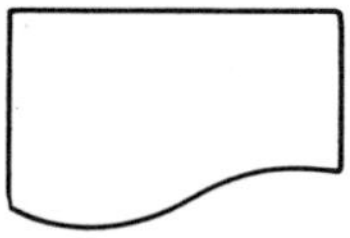

DOCUMENT. A symbol shaped like a torn piece of paper shows that data is to be recorded or stored on a hard copy document, or that the output will be a written, printed, or typed hard copy document. Preparation or storage of reports, signatures, forms, invoices, statements, and quarterly forms are shown using this symbol.

PUNCHED TAPE. A symbol shaped like a piece of paper tape specifies data is stored on or being punched into paper tape. This

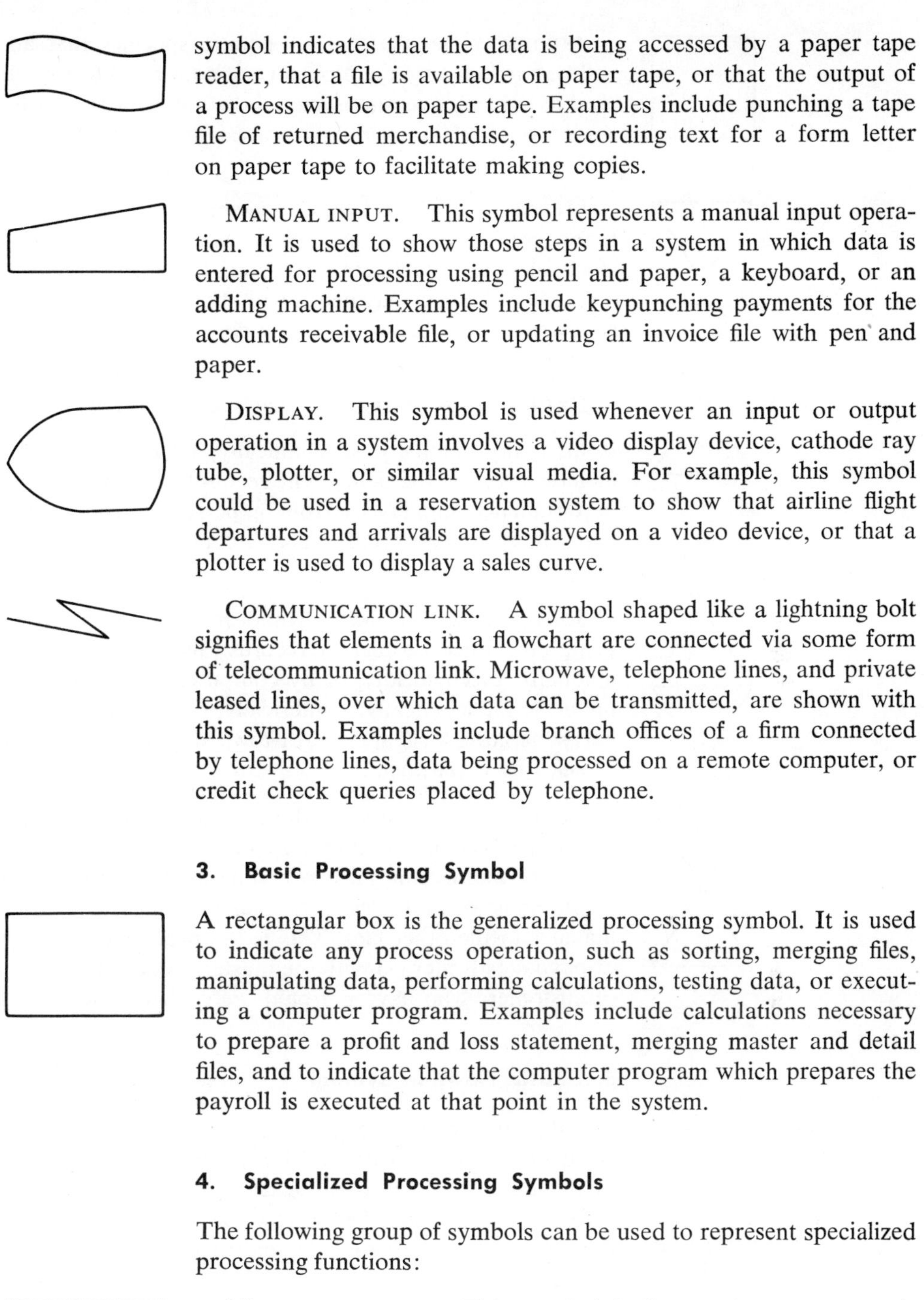

symbol indicates that the data is being accessed by a paper tape reader, that a file is available on paper tape, or that the output of a process will be on paper tape. Examples include punching a tape file of returned merchandise, or recording text for a form letter on paper tape to facilitate making copies.

MANUAL INPUT. This symbol represents a manual input operation. It is used to show those steps in a system in which data is entered for processing using pencil and paper, a keyboard, or an adding machine. Examples include keypunching payments for the accounts receivable file, or updating an invoice file with pen and paper.

DISPLAY. This symbol is used whenever an input or output operation in a system involves a video display device, cathode ray tube, plotter, or similar visual media. For example, this symbol could be used in a reservation system to show that airline flight departures and arrivals are displayed on a video device, or that a plotter is used to display a sales curve.

COMMUNICATION LINK. A symbol shaped like a lightning bolt signifies that elements in a flowchart are connected via some form of telecommunication link. Microwave, telephone lines, and private leased lines, over which data can be transmitted, are shown with this symbol. Examples include branch offices of a firm connected by telephone lines, data being processed on a remote computer, or credit check queries placed by telephone.

### 3. Basic Processing Symbol

A rectangular box is the generalized processing symbol. It is used to indicate any process operation, such as sorting, merging files, manipulating data, performing calculations, testing data, or executing a computer program. Examples include calculations necessary to prepare a profit and loss statement, merging master and detail files, and to indicate that the computer program which prepares the payroll is executed at that point in the system.

### 4. Specialized Processing Symbols

The following group of symbols can be used to represent specialized processing functions:

MANUAL PROCESS. This symbol indicates that a process is being performed manually, without the use of machine or computer. Examples would be filing records, calculating loan information with pen and paper, or taking inventory.

Auxiliary operation. A square specifies that an operation will be carried out on equipment not connected to a computer. It is used to indicate operations such as listing records recorded on magnetic tape or a line printer, or processing data punched into cards with an accounting machine.

Merge. The inverted triangle shows that two or more files are to be combined into one. This symbol can indicate a manual, mechanical, or computer operation. Examples include merging a detail file showing payments wtih the records in the master file showing account balances, or merging a list of new employees into the payroll file.

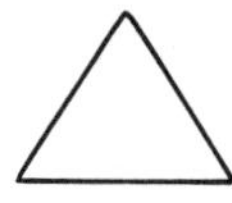

Extract. A triangle indicates data or records being pulled from a file, manually, mechanically, or electronically. Examples include selecting the delinquent accounts in a file, removing names of terminated employees from a personnel file, or marking any items in a list of expenses that are not within the expected cost range.

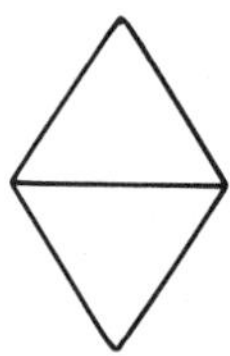

Sort. This specialized symbol indicates a sorting operation is being performed on files or records. The process can be executed manually, mechanically, or by computer. Examples include ordering a file alphabetically, numerically, or by account or stock number; separating local and out-of-town shipments; or sorting a group of sales by department.

Collate. This symbol shows that two or more files are to be combined into one file that may, or may not, be in the same order as any of the original files. A common example of collating would be merging several files into one, pulling aside records without matches in the other files. This is the process followed when a file of payments is merged with the master account file, and the records of customers who have not paid are pulled aside.

### 5. Basic Flowline Symbol

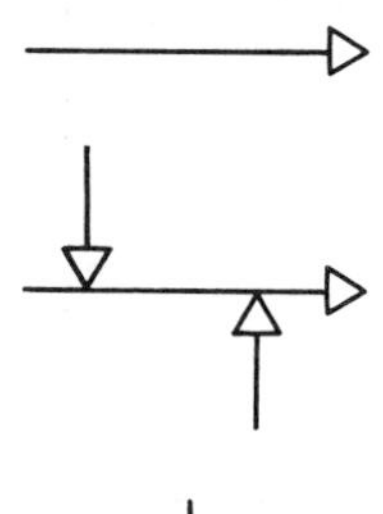

The direction of movement or flow between elements of a system is shown with the flowline symbol—a line that connects both points in the flowchart. An arrowhead at one end of the line indicates the direction of flow. If no arrowhead is shown, it is assumed that flow moves from left to right, or from the top of the page to the bottom.

Arrowheads are used to show where flowlines join, and to indicate their new direction.

Lines that cross each other and continue in separate directions are assumed to be unrelated.

### 6. Connector Symbol

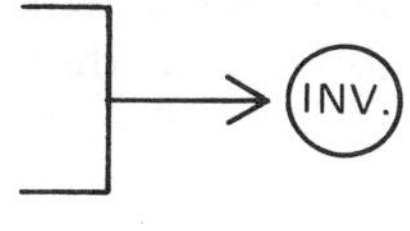


A second way to show flow or movement between elements in a system is with the connector symbol. This is a small circle with the name of the next sequence inside. One symbol is used at the terminal point of a sequence, and a second symbol, with a matching label, points to the beginning of the next sequence where flow resumes. Arrowheads indicate the direction of flow.

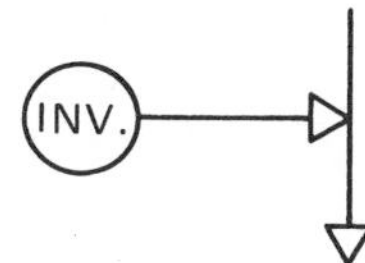


This symbol is often more convenient than flowlines for use in flowcharts. It eliminates lines that cross or form confusing junctions. It is useful for directing flow to sequences on other pages in a multipage flowchart. It improves the appearance and clarity of flowcharts.

### 7. Basic Annotation Symbol

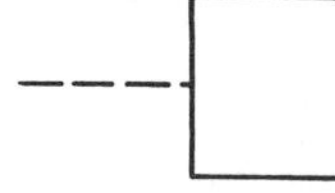

Notes or comments can be added to a flowchart with the annotation symbol. This symbol, an open-ended box, encloses additional explanations or descriptive text related to a given step. A broken line points to the step in the flowchart that it describes.

Five other symbols appear in the ANSI summary (Figure 8.1.). These symbols—the decision, predefined processing, preparation, terminal, and parallel mode symbols—are used primarily in flowcharting computer programs, and are not discussed in this text.

## HOW TO DRAW SYSTEM FLOWCHARTS

The rules and symbols used in the preparation of a system flowchart have become standardized and are easy to understand. They represent an almost universal language for communications between systems analysts.

The care and detail used when preparing a given flowchart varies according to the chart's immediate purpose. For preliminary planning and experimentation, the analyst will probably draw a rather hurriedly prepared chart using only pencil and paper. The symbols are drawn freehand and show only the major elements in a sequence or system. Later, the chart may be expanded, refined, improved, and drawn permanently, using a template, ruler, pen, and ink.

### Physical Considerations

Flowcharts should be drawn on 8½″ × 11″ paper (or other standard page size if required). If the charts are to be kept permanently, a good grade of 20-pound bond paper should be used.

Leave adequate margins (a minimum of 1″ on all sides of the page, 1½″

if binding is a consideration). Leave sufficient space between symbols. The flowchart should be labelled at the top with the name of the system or subsystem to which it relates and other important details, such as the date and revision number. If several pages are involved in a single flowchart, they should be numbered consecutively.

A template should be used to assure that the shape of each symbol adheres to recognized standards. Figure 8.2 shows a plastic template that contains the ANSI standard symbols. It is available from RapiDesign, Inc., Burbank, California, No. 54A. A similar template, with additional symbols, is available from IBM Corporation, White Plains, New York, Form No. X20-8020.

### Content Considerations

These are a few techniques and conventions that should be followed when drawing flowcharts to facilitate tracing the flow and procedures represented in a chart:

Wherever possible, draw the direction of flow from left to right, and from the top of the page to the bottom. In instances where this is difficult or confusing, use arrowheads to indicate the direction of flow.

A symbol may vary in size, but its shape must remain proportional and its orientation on the page unchanged.

Always attempt to show a data flow path in the simplest way, but avoid crossing flowlines whenever possible. The connector symbol should be used to prevent crisscrossing of lines. Insert arrowheads liberally throughout the chart to show the direction of flow.

Many analysts prefer to have flowlines enter a box from the top only. If more than one line is involved, they join them before entry, so that only one line actually goes to the box. Several flowlines may leave a box, either from separate points in the box or from one line which then separates into several branches.

Include descriptive text within each symbol to define the operation represented by the box. Use short, concise terms, and avoid ambiguity. For example, the term "Print Pick List" is better than "Print Report"; and the term "Compute Net Commission" gives more information than "Commission."

Use the annotation, or comment, symbol liberally throughout a flowchart. (In this text, comments have often been deleted from flowcharts since they are described in some detail within the text.) Annotative text should contain important information about data, data paths, processes, reports, media, and work stations that cannot be conveniently included within a symbol.

Several levels of flowcharts are often used to describe a single system. A broad block flowchart may show only the major elements. Other, more detailed charts may include more elements or break out blocks into their smaller components. The level of detail shown on a flowchart should be

Figure 8.2. Flowcharting template

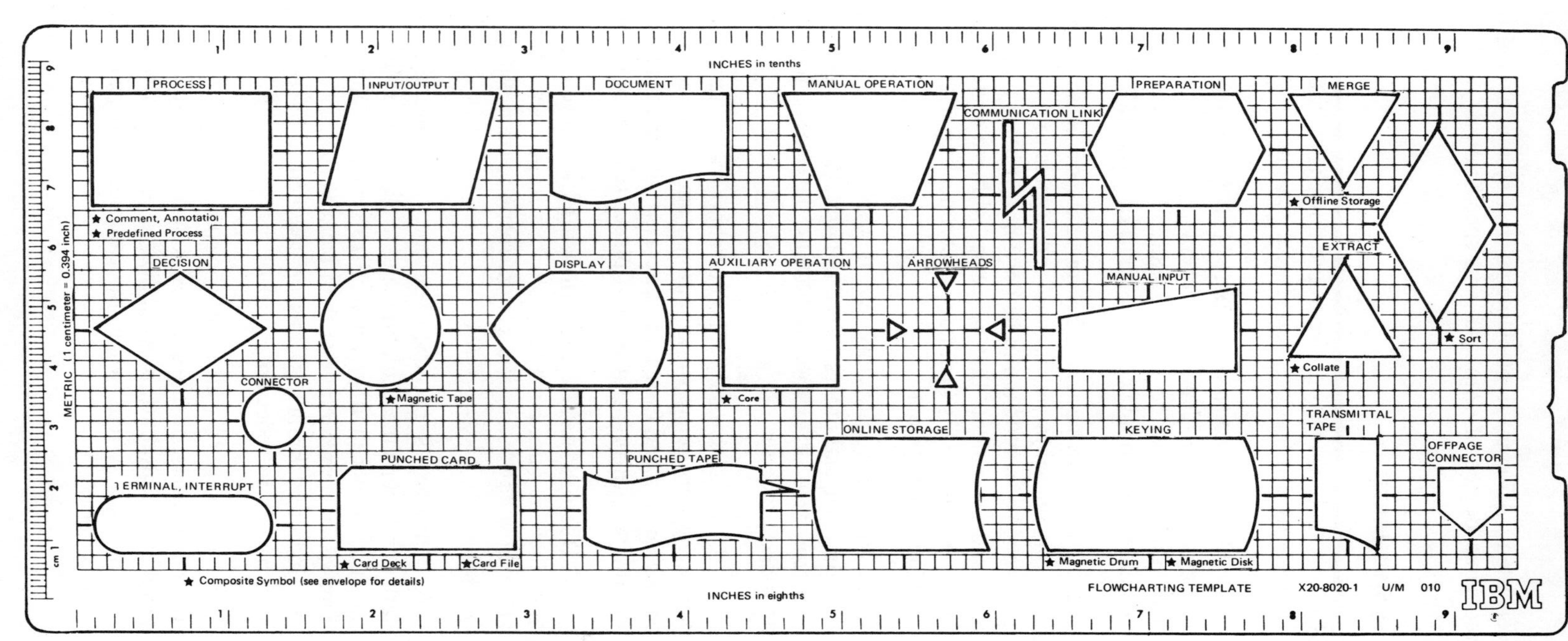

consistent. All equivalent elements in a system should be represented with the same degree of emphasis and specification.

## ADVANTAGES OF FLOWCHARTS

Systems analysts often prefer to use flowcharts, rather than verbal or textual descriptions of a system, for several reasons:

1. COMMUNICATIONS TOOL. The flowchart is a convenient method of communicating ideas and concepts among engineers, designers, management, data processing people, and others. It provides a concrete, visual medium for describing complex plans or problem solutions.

2. PLANNING TOOL. The flowchart is a flexible tool when planning and designing a new system. It enables designers to diagram concisely, on paper, the essence of a system and its flowlines. It helps them visualize the relationships and movements that will exist within a system, while it is still in the planning stages. This facilitates experimenting with different data flow paths and alternative systems.

3. PROVIDES AN OVERVIEW OF A SYSTEM. Flowcharts enable the systems designer and others to see the important elements and relationships in a system, free of extraneous details. It gives a bird's eye view of major steps and processes.

4. DEFINES ROLES. The flowchart spells out in pictorial form the role played by personnel, work stations, and processes in a system. It points out where data and forms are generated; the end results or reports produced; and the people, procedures, and places related to the process.

5. DEMONSTRATES RELATIONSHIPS. Flowcharts point out relationships that may not be readily obvious from a verbal or textual description. They show how parts of a system relate to each other, as well as specifying the sequences of data flow.

6. PROMOTES LOGICAL ACCURACY. Flowcharts require designers to clarify and refine their thinking and to describe explicitly all paths, branches, and movements in a network. Errors in logic are often more easily seen from visual representations than from verbal descriptions. The analyst can spot incorrect branches, missing elements, or perhaps discover a more efficient way to redesign a part of the system.

7. FACILITATE TROUBLESHOOTING. System problems, breakdowns in communications, or unforeseen developments are usually more readily diagnosed if a flowchart is available. It serves as a blueprint of the inner workings of a system, which the troubleshooter can refer to when checking out a system failure.

8. DOCUMENTS A SYSTEM. Flowcharts record graphically, formally, and permanently the important elements of a system. This facilitates making changes, revisions, and modifications at a later date. It enables those unfamiliar with a given system to grasp its major elements easily and quickly.

## DECISION TABLES

The systems analyst has another graphic tool available for use in flowcharting systems. This technique, called a decision table, is of particular value in systems with many options and branches. Decision tables are charts, or tables, that graphically display all possible conditions and related actions in the form of a matrix.

A decision table is composed of several parts (Figure 8.3). The condition stub lists all possible conditions that may be present in a situation. The action stub lists all possible actions that may be taken. The condition entry shows all possible combinations of conditions that may occur, and the action entry indicates what actions should be followed for each combination. The entry sections are lined off in vertical columns called rules. Each rule specifies what actions are to be taken for each set of conditions.

**Figure 8.3.** Part of a decision table

| CONDITION STUB | CONDITION ENTRY |
|---|---|
| ACTION STUB | ACTION ENTRY |

Decision tables serve several purposes. They enable the analyst to lay out a complex variety of conditions and to specify definitively how each set of conditions should be handled. The analyst uses the decision table as an additional aid when flowcharting a system. It may be used to break out a larger block into its smaller components. The rules developed in a decision table can be used as the basis for a procedure manual for employees, to tell them how to handle various situations, and to assure consistent treatment.

Decision tables are also used when flowcharting computer programs. They graphically define all the possible options in a problem and the specific actions that the computer program must execute for each set of conditions.

Decision tables provide several advantages:

1. PREVENT OVERLOOKING IMPORTANT ELEMENTS. The design of the decision table lessens the possibility of omitting essential or related elements.

2. ASSURES CONSISTENCY. Decision tables assure that a given set of conditions will be handled consistently.

3. PROVIDES DOCUMENTATION. Decision tables are an easily understood written statement that others may reference to understand how a system functions.

4. REDUCES COMPLEX SITUATIONS TO THEIR SIMPLEST TERMS. Decision tables force analysts to refine their thinking and to clarify details. The decision table cannot be completed until the analyst has explicitly defined all conditions and actions that are involved.

Figure 8.4 illustrates a decision table used in an order processing system. It defines the various ways to handle orders for items that are low in stock. The systems analyst has listed several conditions that are involved—is it a sale item, is the customer an old or new account, is it a large or small order, and why is the item low in stock. In this example, there are 17 rules, indicating that 17 combinations of conditions can exist.

Five different actions are possible—alone or in combination—in response to each rule. These include filling the order, preparing purchase requisitions, confirming orders, preparing invoices, or cancelling orders.

**Figure 8.4.** Order processing system decision table

| | 1 | 2 | 3 | 4 | 5 | 6 | 7 | 8 | 9 | 10 | 11 | 12 | 13 | 14 | 15 | 16 | 17 |
|---|---|---|---|---|---|---|---|---|---|---|---|---|---|---|---|---|---|
| Supply – adequate | Y | N | N | N | N | N | N | N | N | N | N | N | N | N | N | N | N |
| Supply – below minimum | N | Y | Y | Y | Y | Y | Y | Y | Y | Y | Y | Y | Y | Y | Y | Y | Y |
| Supply – out of stock | N | N | N | N | N | N | N | N | N | N | N | N | N | Y | Y | Y | N |
| Item discontinued | N | N | N | N | N | N | N | N | N | N | N | N | N | N | N | Y | |
| Purchase requisition made | | Y | N | Y | N | Y | N | Y | N | Y | N | Y | N | Y | N | | |
| Old Account | | Y | Y | Y | Y | Y | Y | N | N | N | N | N | N | | | | |
| Large order | | Y | Y | Y | Y | N | N | Y | Y | N | N | N | N | | | | |
| Sale item | | N | N | Y | Y | | | | | Y | Y | N | N | | | | |
| Write pick order | X | | | X | X | X | X | | | | | X | X | | | | X |
| Make purchase requisition | | | X | | X | | X | | X | | X | | X | | X | | |
| Cancel order | | | | | | | | | | | | | | | X | | |
| Write order confirmation | | X | X | | | | | X | X | X | X | | | X | X | | |
| Prepare invoice | X | | | X | X | X | X | | | | | X | X | | | | X |

## EXAMPLES OF SYSTEM FLOWCHARTS

To illustrate how system flowcharting symbols are used, two rather simplified business systems will be described. One is a typical payroll processing system, which produces reports and paychecks. The other is a financial system for a savings and loan bank. In this system, teller terminals located in several branches of the bank are connected to a central computer via ordinary telephone lines.

### Payroll Processing System

Figure 8.5 is a system flowchart of a payroll processing system for a company with several hundred employees. Data for the system comes from two sources—time cards and the master payroll file stored on magnetic tape. The system produces three types of output—an updated master file, a payroll ledger, and the paychecks for the current week.

**Figure 8.5.** Payroll processing system

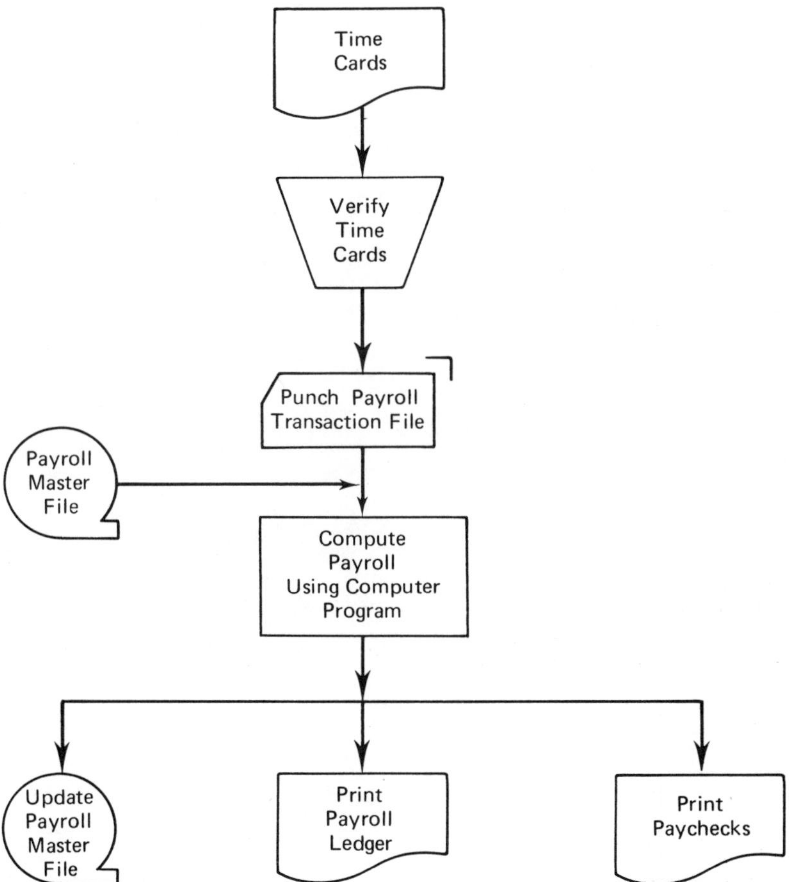

Data for the system originates with the time cards (Figure 8.6). These source documents are punched in and out each day on a time clock to record the number of regular and overtime hours worked by each employee. At the end of each week, the cards are verified by the department foreman and sent to the payroll department for processing. The totals recorded on each time card are transferred onto punched cards by a keypunch operator to form a new file called the payroll transaction file.

The master payroll file, contained on a reel of magnetic tape, contains permanent information on each employee, such as the date of employment, pay rate, overtime rate, accumulated withholding tax contributions, number of exemptions, credit union deductions, and retirement contribution rate.

Once each week the payroll transaction and master files are processed with a computer program to produce the payroll. The computer program performs the necessary mathematical and data manipulation operations to generate an updated master payroll file on magnetic tape. This file becomes the master file that will be used to prepare next week's payroll, and is also used for the preparation of quarterly and yearly tax reports and forms.

The computer also prepares a payroll register. This is a hard copy print-

**Figure 8.6.** Time card

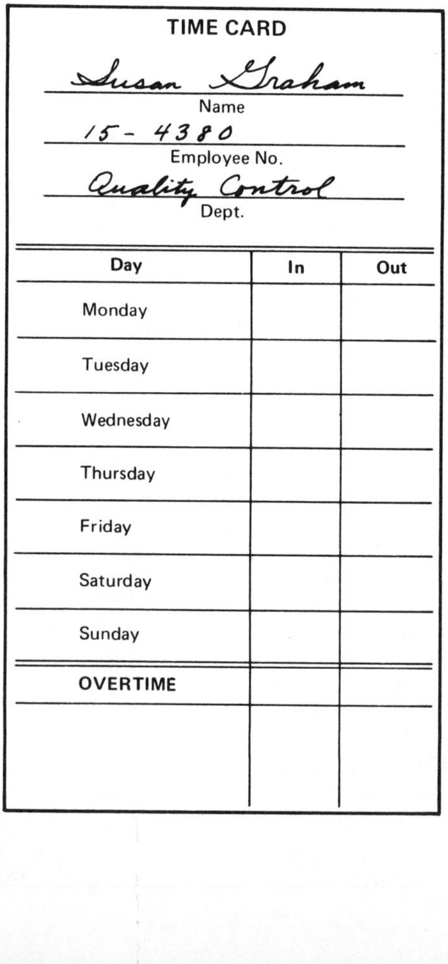

TIME CARD

Susan Graham
Name

15-4380
Employee No.

Quality Control
Dept.

| Day | In | Out |
|---|---|---|
| Monday | | |
| Tuesday | | |
| Wednesday | | |
| Thursday | | |
| Friday | | |
| Saturday | | |
| Sunday | | |
| **OVERTIME** | | |
| | | |

out, listing each employee by name, number, number of hours worked, and other pertinent information such as deductions and net earnings. This hard copy is kept in the management office. Finally, the computer prepares each employee's paycheck, along with a voucher showing gross pay, all deductions, and net pay.

### Savings and Loan Financial System

Figure 8.7 is a typical example of a data flow system used by savings and loan associations to process teller transactions. Data on transactions is entered on remote computer terminals located at each branch and relayed to a central computer via ordinary telephone lines. (See Figure 8.8.) The

**Figure 8.7.** Savings and loan financial system

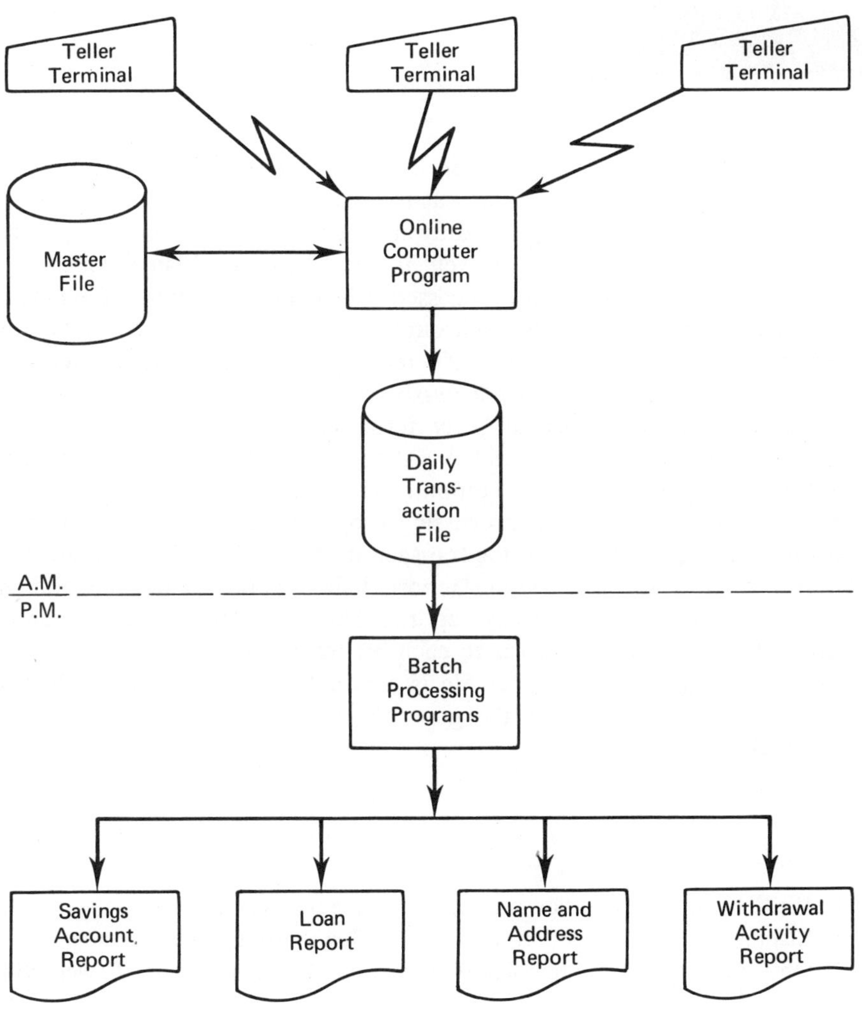

Figure 8.8. Online teller terminal

Courtesy of Burroughs Corp.

information is transmitted directly to the computer at the time of the transaction, while the customer remains at the window.

Information on deposits, withdrawals, name changes, deletions, other file maintenance procedures, as well as data regarding loan payments or new loans, are transmitted and processed. The computer updates the master file, which shows all accounts, their current balances, and related data. Immediate results are returned to each terminal, and the customer receives an immediate update of his or her passbook.

The system is online because each terminal is connected directly to the computer and is served immediately. At the end of the business day, after the branches have closed, the computer is switched to a different type of operation. A group of batch processing programs are executed to prepare various daily reports based on the transactions that occurred during the day. These include a savings account report, loan reports, name and address reports, and withdrawal activity reports. The next morning, a messenger delivers these hard copy reports to each branch manager.

Data recorded on the master file is also used to generate monthly and quarterly special reports for the government, annual earning reports, or personnel reports.

## EXERCISES

1. Write a definition of a flowchart and describe two types.
2. List the names and functions of six major flowcharting symbols.
3. Obtain a flowcharting template (see text) and draw the major flowcharting symbols.

4. Give three general rules regarding physical considerations in preparing flowcharts.
5. What advantages does a template have in preparing flowcharts?
6. Give five content considerations to follow when preparing flowcharts.
7. List six advantages of using flowcharts in problem solving.
8. Describe decision tables and how they are used.
9. Contrast the differences in appearance between decision tables and flowcharts.
10. Draw a system flowchart showing the registration steps followed at your college.
11. Modify the above flowchart to include improvements in the registration process.
12. Study a business organization and prepare a flowchart tracing the major steps in the business system.
13. Select a business activity and prepare a decision table. Show various conditions and actions to be taken.
14. Draw a flowchart of the steps involved in reconciling a monthly bank statement.
15. Prepare a decision table for an order processing system which has a minimum of five rules.

# Chapter 9

# Planning the System

Systems planning is initiated when the systems analyst is called upon to remedy or solve a problem. As in other areas of systems work, the principles of the scientific method are relied upon for guidance and accuracy.

## RECOGNIZING THE PROBLEM

Problems come to the attention of the analyst in several ways. In some organizations, the systems department operates conservatively and waits until difficulties or possible problem situations are brought to its attention. In other organizations the systems department takes an active role in ferreting out and anticipating problems.

Systems people are often measuring, examining, studying, and comparing elements in a system to pinpoint problem situations that are beginning to develop, before they can disrupt the other elements in the system. The analysts attempt to incorporate relevant improvements and techniques as they become available to improve efficiency, increase benefits, and prevent obsolescence of a system.

Very often management will approach the systems department with indications of existing problems, or a request to consider the feasibility of a specific system or installation. Sometimes suggestions or requests for changes or improvements come from personnel in other departments, or from managers or supervisors.

Complaints from dissatisfied customers and criticisms from other employees are often indirect indications of existing problems. Accounting data may point to specific problem areas, such as a cost analysis report that shows unexplained and unexpected increases in expenses or overhead, ratios that show unreasonably high costs in some areas, or excessive overtime or reject rates.

## DEFINING THE PROBLEM

After a problem situation has been recognized, the systems analyst moves on to the next step—the problem must be investigated and analyzed. A brief, quantitative assessment is made of the ways in which the problem manifests itself. This indicates such conditions as the number of hours a report is delayed, percentage of rejects, and numbers of complaints.

Next, the systems analyst determines what goals, outcomes, or improvements would solve or diminish the problem. These are also expressed in quantitative terms: reduction in cost, decrease in turnaround time, increase in fringe benefits.

Now the analyst must consider and recommend ways in which the problem can be solved and the goals reached. This may involve repairing or redesigning an existing system, replacing a system, or in some instances insuring that an established system is thoroughly documented and is being properly operated.

Once these initial steps have been carried out, the analyst considers whether it is practical to make the recommended changes. Will the benefits gained from modifying an existing system, or installing a new one, be worth the costs, effort, and disruption involved?

For extensive, complex changes, this becomes a critical and vital question which cannot be answered lightly or without a thorough analysis of alternatives and outcomes. Firms may have to invest hundreds of man hours and thousands of dollars, and suffer disruption of office routines or deliveries of goods and services to customers.

## FEASIBILITY STUDY

The systems department has developed a sophisticated means of evaluating the practical aspects and details involved in implementing a new or modified system. This analysis is called a feasibility study. Its purpose is to gather, analyze, and document the data needed to make an informed, intelligent decision regarding a system's practicality.

This analysis is done in three phases. The first, called the preliminary study, is concerned with determining whether or not the direct and indirect benefits gained from the new system will be greater than the costs involved. If the answer is no, the feasibility study ends there and the project is temporarily or permanently abandoned.

If the answer is yes, the analysis enters the second phase—the investigative study. There the problem is carefully defined and all details in the solution are specified.

The last phase of the feasibility study is the final report. It fully documents the work done during the first two phases. All expected costs, benefits, and outcomes are shown. It tells how and when the new system should be implemented.

Management will use these recommendations as the basis for decisions regarding the new system. The following chapters discuss the next steps in the process—system installation, evaluation, and documentation.

Some authors view the feasibility study somewhat differently. They separate its functions into two distinct areas—the feasibility study and the system study. Under this arrangement, the feasibility study considers whether a new system can be developed and whether further efforts should be made to investigate it. The study then considers the practical aspects of planning, developing, and implementing the new system. Regardless of the terminology used, the same analysis and development must be undertaken.

### Who Will Conduct the Study?

After the decision has been made on a feasibility study to examine a problem situation or a system modification, the next question is, who will conduct it? There are four ways in which this is usually handled:

1. EMPLOY OUTSIDE CONSULTANTS AND EXPERTS. In this arrangement, a firm that specializes in performing this type of analysis is hired to conduct the study. Personnel working for such firms have skills and training in this area and receive a wide exposure to other, similar situations and solutions.

In some arrangements, the consulting firm performs all phases of the feasibility study and implements the new or modified system. In other instances, it only conducts the study, and the host firm, or other outside experts, manage the implementation.

2. CONTINUING COMMITTEE. Another approach is to have a continuing committee perform the feasibility study. In this arrangement, a group of individuals assumes the responsibility of evaluating a new system's feasibility. The members usually represent many areas of the firm—management, systems analysts, line employees, foremen. In fact, all personnel who will be involved in designing or implementing the system are often included in its membership.

New members may be assigned, as necessary, to maintain a comprehensive and relevant representation. This approach gives the firm an experienced group that constantly monitors company system and data flow. This facilitates spotting and solving problems early, or anticipating situations that should be improved or modified.

3. TASK FORCE. Another arrangement is to have a task force implement a feasibility study. A task force is a group charged with the one-time responsibility of studying a given system problem. Members of the task force may be drawn from management, employees, or systems analysts. After the particular assignment has been completed, the group disbands. If another problem arises later, another task force is set up to solve that specific situation.

4. PROJECT DIRECTOR. Sometimes one person is entrusted with the responsibility of conducting, or directing, a feasibility study. The project director may be a systems analyst on the systems department staff, or may be drawn from management. The director may conduct all the phases of the study alone, or may supervise one or more assistants.

The arrangement a firm selects to carry out a feasibility study depends on several factors—the size of the firm; how soon the new system is to be installed; the anticipated complexity and cost of the new system; and the availability of personnel with the necessary experience and knowledge.

### Phases of the Feasibility Study

The three phases of the feasibility study are consecutive, and the results of one part are used to guide the development of the next. If the results of one phase show that the new system, as designed, would not be feasible, the entire project may be terminated at that point, or the step repeated to investigate an alternative solution.

PRELIMINARY STUDY. The first phase of the feasibility study answers the question, "Should further time, money, and resources be spent in studying or developing a new or modified system?" Costs for developing and operating the system are estimated and compared to the expected direct and indirect benefits. The results of this cost benefit analysis are used to determine whether or not the next phase should be performed.

In order to assess accurately the costs and benefits of a system, certain information must be gathered and documented. The structures and organizations of the present system, the projected system, and the firm itself are studied and analyzed.

The elements and relationships in the existing system must be determined and measured. Time elements involved in performing various actions are calculated. The output is described in quantitative terms. Cost data for operating the system is gathered and allocated. The system's strengths and weaknesses are defined and expressed in quantitative terms.

The objectives and improvements from a new system are also stated in quantitative terms. The study group investigates other firms and systems to see how they handle similar situations and problems. Alternative plans are discussed and one or two selected for further investigation and development. Equipment capabilities are compared, and competitive equipment bids are obtained.

The group estimates the costs and benefits that will result from the development and operation of the new system. This includes costs in planning, developing, and implementing the system: salaries; preparation of computer programs; materials; commissions to outside consultants or experts; purchase, lease, and installation of new equipment; training personnel to operate the new equipment; and any necessary modifications to the building. Conversion costs involved in changing from the old to the new system must also be estimated.

The costs involved in operating the new or modified system are assessed: salaries, forms and other materials, equipment maintenance and replacement, updating employee skills, building maintenance.

A study is made of the benefits that the firm expects to derive from implementing the new system. These include direct benefits such as savings in costs and expenses (salaries, physical space, equipment, equipment operation, personnel training), and improved output and service (reduction in errors, turnaround time). Direct benefits also include the ability to produce comprehensive reports with more detailed interpretations more frequently.

Benefits that accrue from a system can also be indirect—such as increased morale, or greater understanding of the organizational structure of a firm or system. Attitudes may change as well. Increased awareness of the existing situation will sometimes lead to improvements in other departments or systems, and even in the organizational structure of the firm itself. A system designed for one particular problem or department may be "exportable," usable, with little or no modification, in other branches, departments, by other firms, or to solve other problems.

In preparing the cost/benefit analysis, the study group is concerned with questions such as these:

1. What will be the differences in cost and profit between the old and new systems? Is a new system justified on the basis of cost alone?

2. Will a new system increase the firm's output capability? Will the company be able to prepare more letters and correspondence, handle more phone queries, process orders faster, ship goods quicker, or reduce errors in billing?

3. Will the cost of maintaining the new system be less than the old? What will be the true savings in maintenance costs in both the long and short run?

4. What indirect benefits will result from a new system? Will changing the elements of one system improve the operation of another system? Will

improved shipping documents speed inventory preparation, billing, or accounting?

5. Will the new system be more reliable and dependable? Are there reasonable expectations of less down time and fewer periods of inactivity due to breakdowns of equipment and communication failures?

6. What improvements in personnel attitudes will result? Will the new system produce greater motivation, resulting in less absenteeism and greater productivity?

If the cost/benefit analysis suggests that the new or modified system will be feasible, plans are made to continue on to the next phase. The individuals who will conduct this stage are specified. Sometimes outside consultants will be employed at this point to carry out the remainder of the project. Sometimes the same group that conducted the preliminary stage handles the next part, and sometimes it is turned over to the systems department.

The study team usually recommends a time table for planning, developing, and implementing the system, or it may suggest that additional time be allowed for more study.

A report, detailing the work and recommendations of the study, is made for management. Based on this report, the decision is made whether to end the project or begin phase two.

INVESTIGATIVE STUDY. In this phase, the actual specifications, elements, and relationship of the new system are planned in considerable detail. The costs involved in development, implementation, and operation are reviewed and updated to reflect the most recent specifications, estimates, and bids. A detailed time schedule for installing the system is drawn up to serve as a guide during the implementation phase.

Before the group can begin the design work, it may be necessary to make a more thorough analysis of the present situation or the existing system. It may make a more detailed statement of the problem, its causes, and how the new system will meet the expected objectives.

During this examination the study group uses interviews, questionnaires, time and motion studies, direct observation, and the other techniques discussed in previous chapters to gather the necessary information. It considers many factors and pre-existing limitations that must be built into the new system. New systems or modifications must be compatible with systems already in operation within the firm, and must cause as little disruption as possible. The new system must be able to accept data generated by the other systems and output results in a form suitable for their use.

The study team must determine how and where the new and existing systems interact, and what changes are necessary to make them compatible. It indicates which constraints must be designed into the new system, and what elements in the old system will have to be modified.

Existing equipment must be able to perform the type of data processing planned for the new system, or plans must be made to acquire suitable equipment. The degree of personnel retraining that will be necessary must be considered. The availability of employees with the needed skills and experience is often an important factor in systems design.

The study team designs a system geared to meet the objectives and existing parameters. It may consider alternate designs or methods of solving the problem before selecting the one that best meets the firm's needs.

All elements that compose the system are specified. The required equipment is selected, and information regarding exact model numbers, price, purchase and leasing details is indicated. Work stations, job responsibilities, and location of equipment are carefully spelled out, along with salary expectations, training requirements, and availability of suitable personnel. Changes that must be made in the physical plant are specified, along with the expected costs and the anticipated time required to make these changes.

Files, records, forms, and reports used or generated by the system are laid out in detail. The work stations, equipment, and personnel who will be involved in processing this data are indicated.

Information regarding computer programs needed by the system is stated, including costs, who will write them, and a description of the configuration of equipment necessary to execute them. Data input and output generated by the programs are specified. The time required for each operation in the system is estimated and documented.

The study group determines and documents the movement of data, information, activities, and personnel in the new system with system flowcharts. Decision tables are used wherever necessary to clarify situations with many alternative data flow paths.

A major part of the investigative study is preparing the implementation schedule. In this step, the study group recommends how the new system should be installed, and prepares a timetable to guide the transition from old to new as smoothly as possible.

If necessary, a temporary backup system, for use during the changeover, is detailed. Many times the old system continues to operate while the new system is being implemented. The study team recommends the best way to phase out the old system with the least disruption.

Recommendations for preparing the employees for the new system are made. These may include training programs, in-service instruction, meetings, or demonstrations. Costs of any programs and suggested time schedules are given.

The study team also designates who will carry out the implementation of the new system. It indicates who will be in charge of specific areas of responsibility—sometimes the study team itself; at other times the systems department or other individuals in the firm. Sometimes it recommends that outside consultants or vendors be called in to perform specific tasks.

FINAL REPORT. The last step in the feasibility study is the preparation of the final report. This document summarizes the results of the design and planning efforts, and gives the study team's recommendations regarding implementation. Management will use the information in this report as a guide when making future decisions.

The final report itemizes the actual costs involved in the new system, describes the objectives that it will accomplish, and gives a statement of the benefits that the firm will accrue. Based upon these recommendations, management may select one of several courses of action. It may:

1. proceed immediately with the implementation of the new system as outlined in the final report;

2. forestall implementation until a later date when economic conditions improve or other factors change;

3. elect not to implement the new system at all, and terminate consideration of the project.

The course of action can be selected intelligently, since all relevant data is at hand. If the decision is against the new system, management will be aware of the possible direct and indirect benefits that will be lost. If the decision is made to implement the system, the feasibility study has prepared the guidelines to be followed as the firm enters the next phase.

## EVALUATING COMPUTER SYSTEMS

Cost is very often a major factor in deciding whether or not to install a computer system. In fact, the final selection of computer elements in a system nearly always represents a compromise between performance and cost. The major factors considered in this evaluation are discussed below.

### Cost

The purchase price of a computer may vary from under $3,500 for a minicomputer to millions of dollars for a large-scale system. Systems may also be leased on a monthly or yearly basis.

A careful comparison of costs, advantages, and limitations of both approaches should be considered when selecting computer equipment. The final decision must, of course, be made in light of the needs and circumstances of the individual firm.

Outright purchase of a computer may seem to involve less money in the long run, but it brings with it the problems of obsolescence, maintenance, repair, and greater difficulty in modifying the system. Leasing computer equipment increases the flexibility of a system, but the costs may be greater.

### Central Processor Performance

The potential performance of the computer is an important standard of comparison. Speed and capacity are two main measures of performance. They are determined by the length of time required by the CPU to carry out an instruction and the time needed to read or write data in primary storage. These times usually vary from less than one microsecond to several microseconds.

The performance of the CPU is often directly proportional to its cost—faster machines cost more money than do the slower units.

### Hardware Features

Computers differ in construction and the features that are built into them. These factors seriously affect the efficiency and precision with which they can process data, and sometimes even affect the degree of accuracy. Some computers have a wide instruction set built in, allowing them to execute many different types of instructions with a minimum of programming time and effort.

Some machines handle single and double precision, floating point mathematics more efficiently than others. The number of available working registers is also an important factor in writing efficient programs.

### Primary Storage Capacity

CPUs are designed with different primary storage capacities. Minicomputers may have as little as 4,000 bytes (characters or digits) of primary storage, limiting the size of the program or data files that can be processed. Other computers can store many millions of bytes in the CPU at the same time.

Some machines have virtual storage capabilities. This means a machine can move data back and forth between primary and secondary storage at very high speeds. It keeps track of where all data is stored on the secondary storage media with a highly sophisticated indexing system called paging. This allows programs and data files of almost unlimited size to be processed, and it facilitates multiprogramming.

### Software

Computers differ greatly in their software and programming capabilities. The number of computer languages that they can execute and the degree of sophistication of their operating systems are important factors. A computer's ability to run a job in a specific language depends on whether a translating program has been written to convert the coded language instructions into those that the particular machine can execute.

These conversion programs are called compilers, translators, assemblers or interpreters, depending on the type of language they handle. Since machines made by different manufacturers vary in construction, each brand must have its own compiler or conversion program for each language. Some computers can process instructions written only in the Assembler language. Most have several languages available.

The type of operating system written for a particular machine also affects the efficiency of program coding and execution. The operating system is a control program written in the Assembler language by the manufacturer. It comes in various degrees of sophistication. Some are relatively simple, performing only minor input and output processes. Some machines, designed to perform only one type of process, may have no operating system available at all.

Other operating systems can handle multiprogramming. They allow a computer to process several jobs at the same time. Data for one or more jobs will be input and output while the CPU is executing or compiling others. These operating systems will usually allow a higher priority job to be executed ahead of lower priority ones. Other operating systems can perform time sharing functions, allowing several computer terminals to be connected to the CPU at one time.

Another important aspect of software consideration is the library of miscellaneous programs available for a system. This includes utility programs written by the manufacturer to perform routine data processing procedures, such as merging, sorting, and searching files. Other programs include statistical procedures, mathematical routines, games, and demonstration programs. Some machines have extensive libraries available. Others have a very limited one, or none at all.

### Peripheral Equipment and Secondary Storage Devices

The secondary storage and input and output devices needed by a system are important factors in equipment selection. There are two major considerations involved. One is that these devices have different functions, and the appropriate devices must be selected to handle the particular storage or input and output format required by each system. Video terminals should not be selected when hard copy output is needed. Paper tape units will not be adequate when very large quantities of data must be stored.

The second consideration is that not all computers can support all peripheral and storage devices. Both hardware and software elements are involved. The type and number of the devices that can be operated by a particular computer must be considered carefully when comparing different systems.

A group of tables are shown on the following pages to illustrate the type of extensive cost and feature comparisons that are made in the preliminary and investigative studies. The tables include data on a variety of current pieces of hardware and software. They are not intended to be complete descriptions of all machines on the market, but only to represent the kinds of data that concerns the analyst when planning decisions are involved.

Figures 9.1, 9.2, and 9.3 are comparison charts for different types of dictating equipment. They show the costs and features of portable, desk top, and centralized dictation systems made by various manufacturers. Figure 9.4 compares different word processing machines. Figure 9.5 gives the specifications of the major copying machines on the market.

**Figure 9.1.** Portable dictating equipment comparison chart

# 1973 PORTABLE DICTATING EQUIPMENT SPECIFICATIONS CHART

• Standard
+ Optional

| Manufacturer | Reader service number (write on card) | Series or model | Purchase price (nearest dollar) | Weight | Measurements (height X width X depth) | Power source—EO=electrical outlet, AL=auto lighter, B=batteries | Recording media—B=belt, C=cassette, D=disc, T=tape, MB=magnetic belt | Recording time (minutes) | AC adapter | Sensitivity control | Volume control | Review | End of record signal | Phone recording attachment | Battery recharging circuit | Low battery warning | Footnotes |
|---|---|---|---|---|---|---|---|---|---|---|---|---|---|---|---|---|---|
| | | | | | | | | | FEATURES | | | | | | | | |
| Amer. Geloso Electronics | 365 | RV4/10 | 241 | 12 | 5X12X11 | EO | MT | 600 | • | • | • | • | | • | | | |
| | | 4/10 | 230 | 12 | 5X12X11 | EO | MT | 600 | • | • | • | • | | • | | | |
| Craig Corp. | 366 | 2605 | | 2 | 3X5X1 1/2 | EO, B | C | 60 | | | • | | | + | | | 2 |
| DeJur Grundig | 367 | Mark VIII | 104 | 1 | | B | MB | 45 | | | • | | | | | | |
| Dictaphone Corp. | 368 | 10 | 99 | .7 | | B | 13 | 15 | | | | | • | • | | | |
| | | 850 | 415 | 3 | | EO, B | B | 15 | | • | • | | • | • | • | | |
| | | 848 | 149 | 1 | | EO, B | C | 60 | | • | • | | • | • | | • | |
| Dictran International | 369 | Doro 701 | 125 | 2 | | B | C | 120 | | | • | • | | | | • | |
| IBM, Office Products | 371 | Portable | 440 | 1.9 | | B | MB | 10/20 | | | • | • | • | • | | | |
| Karl Heitz, Inc. | 370 | 303A | 230 | 1.8 | | B | T | 24 | | • | • | | • | + | | | |
| | | 300A | 230 | 1.8 | | B | T | 48 | | • | • | | • | + | | | |
| | | 303 | 200 | 1.8 | | B | T | 24 | | • | • | | | + | | | |
| | | 300 | 200 | 1.8 | | B | T | 48 | | • | • | | | + | | | |
| | | 101/1015 | 100 | 1.8 | | B | T | 48 | | + | • | | | + | | | |
| | | Transcriber 302 | 260 | | | EO | C, T | 60-120 | | | | | | | | | 6 |
| | | 301 | 280 | | | B | C, T· | 60-120 | | • | • | | • | + | | | 7 |
| Lanier Business Prod. | 372 | 1300 | 229 | 2 | | EO, AL, B | C | 60 | • | | • | • | • | • | • | • | |
| | | 1100 | 139 | 1.5 | | EO, AL, B | C | 60 | • | | • | • | • | + | • | • | |
| Memocord | 373 | K70 | 179 | .7 | 5 3/4x3X1 | B | C | 90 | | • | • | • | • | + | + | • | |
| Panasonic Inc. | 374 | RV2400S | 79 | 1.7 | 4X6 1/2X3/4 | EO, B | C | 30-60-90 | • | • | • | • | 3 | • | • | • | |
| Phillips Business Sys. | 375 | Norelco 88 | 250 | 1.2 | 7X2 3/4x1 1/2 | EO, B | MT, MC | 30 | + | • | • | 1 | • | + | | • | 4 5 |
| | | Norelco 95 | 160 | .8 | 4 3/4X2 5/8X7/8 | B | MT, MC | 30 | | • | • | 1 | • | | | • | |
| | | Norelco 85 | 100 | .8 | 4 3/4X2 5/8X1 3/8 | B | MT, MC | 30 | | | • | 1 | | + | | • | |

1 - Unlimited. 2 - Built-in condenser mic; digital counter; battery recharging circuit. 3 - Automatic stop. 4 - Includes full indexing capability and soft leather carrying case. 5 - Optional conference microphone. 6 - Has foot pedal controls. 7 - Has mike recorder controls.

**Figure 9.2.** Desk-top dictating equipment comparison chart

# 1973 DESK-TOP DICTATING EQUIPMENT SPECIFICATIONS CHART

• Standard
+ Optional

| Manufacturer | Reader service number (write on card) | Series or model | Purchase price (nearest dollar) | Weight (nearest pound) | Recording media—B=belt, C-cassette, D=disc, T=tape, MB=magnetic belt | Recording time (minutes) | Position of controls—on M=machine, MP=microphone, FP=footpedal, or VA=voice activated | Plays back through M=microphone, R=recorder, or E=earset | FEATURES: Automatic erase | Conference pick-up | Automatic volume control | Measured review | End of record signal | Automatic last word repeat | Phone recording | Footnotes |
|---|---|---|---|---|---|---|---|---|---|---|---|---|---|---|---|---|
| Craig Corporation | 346 | 2706 | | 8 | C | 60 | M, MP, FP | M, E | • | • | • | • | • | | • | 4 |
| | | 2702 | | 6 | C | 60 | M, MP, FP | E | • | | • | • | | | + | 5 |
| DeJur Grundig | 347 | Mark V | 364 | 7 | MB | 45 | MP | M | • | • | • | • | • | | • | |
| | | Executivematic | 294 | 10 | MB | 45 | MP | M | • | • | • | • | • | | • | |
| | | Versatile V | 154 | 4 | MB | | M | E | • | | | | | | | |
| Dictaphone Corp. | 348 | 852 | 525 | 8 | B | 15 | FP | E | | | | • | | • | | |
| | | 851 | 525 | 8 | B | 15 | MP | M, R | | • | • | • | • | | + | |
| | | 812 | 475 | 8 | MB | 12 | FP | E | • | | | • | | • | | 6 |
| | | 811 | 475 | 8 | MB | 12 | MP | M, R | • | • | • | • | • | | + | |
| | | 11 | 275 | 3 | **10** | 30 | FP | E | • | | | • | | • | | |
| | | 2413 | 475 | | C | 60 | M, MP, FP | M, R, E | • | • | • | • | • | • | • | 7 |
| | | 2412 | 425 | | C | 60 | M, FP | R, E | • | | | | • | • | | 7 |
| | | 241 | 425 | | C | 60 | M, MP | M, R | • | • | • | • | • | | • | 7 |
| Dictran International | 349 | Doro 702 | 295 | 15 | C | 120 | M, MP, FP | M, R, E | • | • | • | | | | • | |
| Ford Industries Inc. | 350 | Code-A-Phone 130 | 395 | 16 | T | | M, FP | E, R | • | | • | • | • | + | | 6 |
| IBM, Office Products | 351 | Executary Micro-phone Input | 520 | 9.6 | MB | 10/20 | MP | M, R | • | • | • | • | • | | • | |
| | | Executary Transcriber | 520 | 8.3 | MB | 10/20 | M, FP | E | • | | | • | • | • | | |
| | | Transcriber | 430 | 9.4 | MB | 10/20 | M, FP | E | | | | • | • | • | | |
| | | Micro-phone Input | 395 | 8.6 | MB | 10/20 | MP, R | M | • | | | • | • | | • | |
| Lanier Business Prod. | 352 | Edisette 1977 | 349 | 9 | C | 60 | MP | M, R, E | • | • | • | | • | | + | |
| | | Stenocord | 330 | 9 | MB | 12 | MP | M, R, E | • | • | • | • | • | | + | |
| | | Gray MD-1A | 395 | 11 | MB | 14 | MP | M, R, E | • | • | • | • | • | • | • | |
| Memocord | 353 | AW90DL | 339 | 6 | C | 90 | M, MP | M, R, E | • | • | • | + | • | + | • | |
| | | AW100AR | 299 | 6 | C | 90 | M, FP | E | | | | • | | • | | |
| Miles Reprouducer | 354 | CC | 539 | 5 | B | 180 | M | R, E | | • | • | • | • | | • | 3 |
| Olympia USA, Inc. | 355 | DG20C | 325 | 10 | MD | 10 | MP, FP | M, E | • | • | • | • | • | | • | 2 |
| | | DG21 | 299 | 10 | MD | 10 | FP | E | | | • | • | | | | |
| | | DG20D | 299 | 10 | MD | 10 | MP | M, E | • | • | • | • | • | | • | 2 |
| Panasonic Inc. | 356 | RV2500S | 169 | 9 | C | 30-60-90 | MP, FP | M, E | • | • | • | • | 8 | | • | 9 |
| Phillips Business Sys. | 357 | Norelco 98 | 450 | 7 | MT, MC | 30 | M, MP, FP | M, E, R | • | • | • | 11 | • | | + | 1 3 |
| | | Norelco 96 | 360 | 5 | MT, MC | 30 | M, MP, FP | M, E, R | • | • | • | 11 | • | | + | 3 12 |
| | | Norelco 86 Transcriber | 285 | 4 | MT, MC | 30 | M, FP | E, R | • | | • | 11 | | | | |

1 - Built-in executive-secretary intercom. 2 - Simultaneous dictate/transcribe. 3 - Built-in indexer. 4 - Full microphone controls; synchronized digital counter; automatic telephone recording; Quick-erase during rewind. 5 - Synchronized digital counter; Quick-erase during rewind. 6 - Transcriber. 7 - Incorporates advance Auto-Scan (tm) index system. 8 - Automatic Stop. 9 - Combination dictation/transcribing unit. 10 - Miniature cassette. 11 - Unlimited. 12 - Optional executive/secretary intercom.

**Figure 9.3.** Centralized dictating system comparison chart

● Standard
* Optional

| | | | | | | FEATURES | | | | | | | | |
|---|---|---|---|---|---|---|---|---|---|---|---|---|---|---|
| Manufacturer | Series or model | Purchase price (nearest dollar) | Recording media—B-belt, C-cassette, D-disc, T-tape, MB-magnetic belt | Recording time (minutes) | Plays back through—M-microphone, R-Recorder, T-telephone | Automatic erase | Conference pick-up | Automatic volume control | Measured review | End of record signal | Automatic last word repeat | Phone recording | Simultaneous dictate/transcribe | Number of remote dictating units per central recorder |
| Dictophone Corp. | 187T | 1230 | MT | 180 | T | ● | * | ● | ● | ● | | ● | ● | 2 |
| | 185T | 1585 | MT | 180 | * | ● | * | ● | ● | ● | | ● | ● | 2 |
| | 185 | 1470 | MT | 180 | * | ● | * | ● | ● | ● | | ● | ● | 2 |
| | 181T | 1230 | MT | 180 | * | ● | * | ● | ● | ● | | * | ● | 2 |
| | 181 | 1115 | MT | 180 | T | ● | * | ● | ● | ● | | * | ● | 2 |
| | 180 | 925 | MT | 180 | T | ● | * | ● | ● | ● | | ● | ● | 2 |
| | 707 | 1425 | MB | 12–15 | M | ● | ● | ● | ● | ● | ● | ● | ● | 2 |
| | 705 | 1015 | MB | 12–15 | M | ● | ● | ● | ● | ● | ● | ● | ● | 2 |
| | 190 | 895 | T | 60 | M | ● | * | ● | ● | ● | ● | * | ● | 6 |
| Dictran International | Doro 320 CDS | 745 | C | 60 | M, T | ● | | ● | | ● | | ● | | 2 |
| Ford Industries Inc. | Code-A-Phone 800 | 995 | T | 120 | R | ● | ● | ● | ● | ● | | ● | | 6 |
| IBM Office Products | Micro-phone Input | 568 | MB | 10/20 | M | ● | | | ● | ● | | | | 8 |
| | Dial Input | 1200 | MB | 10/20 | T | ● | | | ● | ● | | ● | | 2 |
| | Tone Input | 2400 | MB | 10/20 | T | ● | | | ● | ● | | ● | | 2 |
| Lanier Business Prod. | 101 | 1330 | T | 400 | T | ● | * | ● | | ● | ● | * | ● | 2 |
| | 102 | 1130 | T | 100 | T | ● | * | ● | | ● | ● | * | ● | 2 |
| | 108 | 1095 | T | 400 | T | ● | * | ● | | ● | ● | * | ● | 2 |
| | LPN | 795 | T | 100 | T | ● | * | ● | | ● | ● | * | ● | 2 |
| | Tele-Edisette | 1330 | C | 360 | T | ● | * | ● | | ● | | * | | 2 |
| Memocord | TDS100 | 995 | C | 90 | M, R, T | ● | ● | ● | * | ● | | ● | | 2 |
| Phillips Business Sys. | 0246 | 1395 | MT, MC | 30 | T | ● | | ● | ● | ● | | ● | | 2 |
| | 0245 | 895 | MT, MC | 30 | T | ● | | ● | ● | | | | | 10 |

2—Unlimited. 3—PBX and/or private wire. 4—Telephone answering. 5—Price for recorder only; other components of system additional. 6—Voice activated. 7—Connects to and utilizes existing telephone system.

**Figure 9.4.** Word processing machine comparison chart

# 1973 WORD PROCESSING SPECIFICATIONS CHART

| For more information about a company's products, write the company's number on the reader service card. For immediate information, each company's phone number and contact is listed. MANUFACTURER AND CONTACT | Reader service number | ● = Standard † = Optional Model name or number | Purchase price: base model or system | Monthly lease or rental: base model or system | Input: enter, correct, and/or edit data via: | Stores information on: | Storage capacity in characters, words, or pages | How to locate stored material | Number of storage stations (tapes, cassettes, etc.) | Type of output | Writing speed (wpm) | SHARED SYSTEMS: Satellite input (yes-no) No. of input stations to CPU | Standard operator training time with purchase or installation |
|---|---|---|---|---|---|---|---|---|---|---|---|---|---|
| IBM Office Products Division Fred Steinberg, info. mgr. 201/848-1900 | 338 | Mag. Card Selectric® Typewriter | 7875 | 185 | Typewriter keyboard | Magnetic cards | 5,000 char.kk | Line Selector | 1 | Selectric® Typewriter | 150 | †r | 2½-days |
| | | Mag. Card Executive Typewriter | 9000 | 235 | Typewriter keyboard | Magnetic cards | 5,000 char.kk | Line Selector | 1 | Selectric® Typewriter | 150 | | 2½-days |
| | | Mag. Tape Selectric Typewriter | 7150 | 195 | Typewriter keyboard | Magnetic tape | 28,800 char.ll | Dial code | 1-2 | Selectric® Typewriter | 150 | | 3 days |
| | | Mag. Card II R Selectric Typewriter | 11,800 | 260 | Typewriter keyboard | Mag. cards & electronic memory | mm | Instant access | 1 | Selectric® Typewriter | 150y | | |
| Redactron Corporation Anthony Mauro, VP 516/543-8700 | 348 | R-4C Single Card | 6400 | 190 | Type-writer | Mag cards | 10,240 char.kk | Unit record | 1 | Heavy-duty Selectric® | 185 | | Unlimited |
| | | R-5C Dual-Card | 8200 | 275 | Type-writer | Mag cards | 10,240 char.kk | Unit record | 2 | Heavy-duty Selectric® | 185 | | Unlimited |
| | | R-4T Single Tape | 6400 | 185 | Type-writer | Mag Tape | 60-70,000 char.ll | numerical code | 1 | Heavy-duty Selectric® | 185 | | Unlimited |
| | | R-5T Dual Tape | 8000 | 265 | Type-writer | Mag Tape | 60-70,000 char.ll | numerical code | 2 | Heavy-duty Selectric® | 185 | | Unlimited |
| | | Power Typewriter | 4995 | 165 | Type-writer | Mag. cards | 10,240 char.kk | Unit record | 1 | Heavy-duty Selectric® | 185 | | Unlimited |
| Savin Business Machines Corporation Gabriel S. Carlin, exec. vice president 914/769-9500 | 349 | Savin 900 Word | 3,995x | 95-175ff,x | Selectric® Type-writerx | Cassette | 12-24 pages | Record display window | 1 | Selectric® Type-writerx | 186 | | 1 day |
| Scribona of North America, Inc. Roy D. Witte, president 301/797-5672 | 350 | 500 | 10,500 | 375 | Type-writer keyboard | Magnetic cartridge | 125,000 char. | vv | 1 | Selectric® Type-writer | 183 | | Yes |
| | | 100 | 7,500 | 250 | Type-writer keyboard | Electronic memory | 5,000 char. | Dial or keyboard control | 1-2 | Selectric® type-writer | 183 | | Yes |
| Singer Graphic Systems George W. Kovatch mktg. admin. 415/357-6800 | 351 | 9400 Editing Terminal | 17,300ad | N/A | Paper-tape & keyboard CRT | Paper tape | 5121 10,240 | Cursor & Scrolling | N/A | Paper tape | 70 cps | ● 3-4 | 2-4 hours |
| Sperry-Remington | 352 | MT 100 | 7200 | 194ww | Type-writer | Cassette | 60,000 char. | Auto search | 1 | Type-writer | 175 | | Yes |

**Figure 9.5.** Copy machine comparison chart

| MANUFACTURER | Reader service number | Model Number | Reprographic process (electrofax, electrostatic) | C-Console T-Tabletop | Machine dimensions (H x W x D, to nearest inch) | Cr-Conveyor F-Flatbed | Price (purchase) | Monthly rental or lease | Cost per 8 1/2 x 11 copy (based on how many copies) | Maximum size of originals | Time required for first copy (in seconds) | Copies per minute (speed) | Roll fed or sheet fed | Makes duplicating masters | Makes transparencies or translucents | Cuts copies to original size | Copies bound volumes | Exposure control (M-Manual A-Auto) | Accepts colored originals | Accepts half-tones |
|---|---|---|---|---|---|---|---|---|---|---|---|---|---|---|---|---|---|---|---|---|
| IBM | 306 | IBM Copier | Transfer Electrostatic | C | 40x35x25 | Fx | 13,000 | v | v | 14x17 | 15 | 10 | R | • | • | | • | M | • | • |
| | | IBM Copier II | Transfer Electrostatic | C | 43x41x28 | F | 22,000 | w | w | No limit | 6 | 25 | R | • | | | • | M | • | • |
| 3 M Company | 307 | "VHS" | Magne-Dynamic | C | 38x68x26 | F | 13,250 | 50pp | .002-.038 | 8 1/2x14 | 3.5 | 20 | S | • | | | • | M | • | • |
| Multigraphics Div. Addressograph-Multigraph | 308 | 6000 | Transfer electro-photography | C | 39x47x30 | F | 8900 | 230 o | .006 | 10 1/2 x 14 1/4 | 7 | 10 | S | | • | | • | M | • | • |
| Royfax | 309 | Royal Bond Copier | Transfer electrostatic | C | 33x38x24 | F | 7995 | 194 b | .026 c | 10x14 | 15 | 10 | S | • | • | | • | M | • | • |
| Savin Business Machines | 310 | 300 | Electrostatic | C | 36x50x22 | C | 4950 | h | .01 | 8 1/2x14 | 10 | 17 | R | • | • | • | • | M | • | • |
| Saxon Business Products | 311 | PPC-1 | Electrostatic | T | 15x54x24 | F | 4495 | 175 | .019gg | 11x17 | 12 | 30 ac | S | | | | • | A | • | • |
| Sperry-Remington | 312 | 530 | Electrostatic | C | | C | 3250 | | | 8 1/2x14 | 8 | 10 | S | • | • | | | M | • | • |
| Van Dyk Research | 313 | Van Dyk 4000 | Xerographic | C | 40x30x68 | F | | 600 | .005-.037 | 14x17 | 7 | 67 | R | • | | • | • | | • | • |
| Xerox | 314 | 660-I | Electrostatic | T | 18x20x36 | C | 3600 | 37 | .0395 | 8 1/2 x 13 1/2 | 25 | 5.5 | S | | • | | | M | • | • |
| | | 914 | Electrostatic | C | 42x45x46 | F | 4650 | 40 | .0425 | 9x14 | 33 | 7 | S | • | • | | • | A | • | • |
| | | 3100 | Electrostatic | T | 37x32x28 | F | N/A | 125 | .0395r | 8 1/2x14 | 8 | 20 | S | • | • | | • | A | • | • |
| | | 720-I | Electrostatic | C | 42x45x46 | F | 8600 | 75 | .0425 | 9x14 | 16 | 12 | S | • | • | | • | A | • | • |
| | | 1000 | Electrostatic | C | 42x45x46 | F | 9800 | 175 | .020s | 9x14 | 20 | 15 | S | • | • | | • | A | • | • |
| | | 4000 | Electrostatic | C | 43x34x27 | F | N/A | 175 | .0395 | 8 1/2x14 | 7 | 45 | S | • | • | | • | A | • | • |
| | | 2400 | Electrostatic | C | 46x67x31 | F | 22,250 | 300 | .025 t | 8 1/2 x 13 1/2 | 9 | 40 | S | • | • | | • | A | • | • |
| | | 3600-I | Electrostatic | C | 46x67x31 | F | 34,150 | 500 | .02 u | 8 1/2x14 | 8 | 60 | S | • | • | | • | A | • | • |
| | | 7000 | Electrostatic | C | 46x67x31 | F | 38,300 | 600 | .006 s | | uu 8 | 60 | S | • | • | | • | A | • | • |

**ac** - 15 - 30 CPM **b** - 4000 copies **c** - above minimum **gg** - 10,000 copies **h** - $187 - 4000 copies; .025 for 4001 to 10,000; .020 for 10,001 to 20,000; .010 over 20,000 **m** - plus conveyor **pp** - minimum $300 including copies **r** - 3,000 copies **s** - 11+ from each original **t** - 4 x 10 from each original **u** - 24,000 copies **uu** - 0, 15%, 28%, 35%, 38% - pushbutton selected **v** - rental $192/mo. including 4000 copies; Other copies .023 to maximum $575. Lease $192/mo. with 4900 copies; other copies .0207 to maximum $518. **w** - rental $295/mo. with 700 copies; other copies .025 to maximum $925. Lease $295 with 8300 copies; other copies .0225 to maximum $833. **x** - moving.

Reprinted from the August 1973 issue of *Modern Office Procedures* and copyrighted 1973 by Industrial Publishing Company, Division Pittway Corporation.

**Figure 9.6.** Large scale computer system comparison chart—Sperry Univac Computer System

**Computer Characteristics** *(continued)*

| | | | Central Processor | | | | | | | | | | | | | | | Readers | | |
|---|---|---|---|---|---|---|---|---|---|---|---|---|---|---|---|---|---|---|---|---|
| | | | Speed (usec) | | Hardware Features | | | | | | Storage | | | | | | | | | |
| Manufacturer/ Model No. | Avg. Purchase Price ($1,000) | Avg. Rental Price ($/Mo.) | CPU Cycle Time | Add Time* | Automatic Interrupt | Floating-Point Arith. | Memory Protection | Indirect Addressing | Editing Instructions | No. Index Registers | Medium** | Maximum Capacity (in characters) | Access Time (in usec) | Typewriter Console | No. CPU I/O Channels | Data Word Length*** | Buffering | Data Collection | MICR | OCR |
| **Sperry Univac Computer System – Circle 19** | | | | | | | | | | | | | | | | | | | | |
| 90/60 | 700 | 16K | 0.6 | 11.7 | x | x | x | x | x | 32 | IC<br>DI | 512K | 27ms | x | 11 | 32 | x | x | | x |
| 90/70 | 1,000 | 25K | 0.6 | 20 | x | x | x | x | x | 32 | IC<br>DI | 648K | 27ms | x | 22 | 32 | x | | | x |
| 418-III | 750 | 18.7K | 0.75 | 8.5 | x | x | x | | | 8 | C<br>DR<br>DI | 393K<br>132m<br>58m | 0.75<br>92ms<br>60ms | x | 8-32 | B | x | x | | |
| 494 | 1,900 | 45K | | | x | x | x | x | x | 14 | C<br>DR | 1m<br>1b | 0.75<br>92ms | x | 24 | B | x | x | x | x |
| 1106 | 1,500 | 32K | 1.5 | | x | x | x | x | x | 128 | C<br>DR<br>DI | 262K<br>1,573K<br>198m | 0.75<br>4.25ms<br>92ms | x | 4-16 | B | x | | | |
| 1108 II | 2,880 | 60K | 0.75 | | x | x | x | x | x | 15 | C<br>DR | 1m<br>1b | 0.75<br>92ms | x | 16 | B | x | x | x | x |
| 1110 | 2,000 | 44K | 0.12 | 1.5 | x | x | x | x | x | 16 | C | | 0.32 | x | 96 | 96 | x | x | | x |
| 9200 | 57 | 1.5K | 1.2 | 120 | x | | | | x | 16 | PW<br>DI | 16K<br>12.8m | 1.2<br>132ms | x | 11 | 8 | x | | | x |
| 9200 II | 129 | 3,475 | 1.2 | 120 | x | | | | x | 16 | PW<br>DI | 32K<br>57m | 1.2<br>75ms | x | 12 | 8 | x | | | x |
| 9210 | 52 | 1.6K | 1.2 | 120 | x | | | | x | 16 | PW<br>DI | 16K | 1.2<br>132ms | x | 11 | 8 | x | | | x |
| 9211 | 75 | 2K | | 120 | x | | | | x | 16 | PW<br>DI | 16K | 1.2<br>75ms | x | 12 | 8 | x | | | x |
| 9214 | 105 | 3K | | 120 | x | | | | x | 16 | PW<br>DI | 16K | 1.2<br>75ms | x | 12 | 8 | x | | | x |
| 9311 | 111 | 3K | 0.6 | 60 | x | | | | x | 16 | PW<br>DI | 16K | 0.6<br>75ms | x | 12 | 8 | x | | | x |
| 9314 | 131 | 4K | 0.6 | 60 | x | | | | | 16 | PW<br>DI | 16K | 0.6<br>75ms | x | 12 | 8 | x | | | x |
| 9300 | 140 | 3.7K | 0.6 | 60 | x | | | | x | 16 | PW<br>DI | 32K<br>12.8m | 0.6<br>132ms | x | 4<br>11 | 8 | x | | | x |
| 9300 II | 151 | 4,090 | 0.6 | 60 | x | | | | x | 16 | C<br>DI | 32K<br>58m | 0.6<br>75ms | x | 12 | 8 | x | | | x |
| 9400 | 380 | 10K | 0.6 | 22.2 | x | | x | | x | 32 | TF<br>DI | 131K<br>58m | 0.6<br>75ms | x | 10 | 16 | x | | | x |
| 9480 | 350 | 85K | 0.6 | 24.6 | x | | x | | x | 32 | IC<br>DI | 262K | 30ms | x | 10 | 16 | x | x | | x |
| **Telefunken Computer GmbH – Circle 20** | | | | | | | | | | | | | | | | | | | | |
| TR 440 | 2,000 | 30K | 0.625 | 2.8 | x | x | x | x | x | | C<br>DR<br>DI | | 0.3 | x | 52 | B | x | | | |
| **Varian Data Machines – Circle 21** | | | | | | | | | | | | | | | | | | | | |
| V72 | 43.2 | | 0.66 | 12.8 | x | | x | x | x | 2 | C<br>DI | 64K<br>186.8m | 10ms | x | 48 | B | x | x | x | |
| V73 | 75 | | 0.66 | 12.8 | x | | x | x | x | 2 | C<br>DI | 32K<br>186.8m | 10ms | x | 48 | B | x | x | x | |
| V74 | 115 | | 0.66 | 12.8 | x | | x | x | x | 2 | C<br>DI | 512K<br>186.8m | 10ms | x | 48 | B | x | x | x | |
| **Xerox Corp. – Circle 22** | | | | | | | | | | | | | | | | | | | | |
| Sigma 3 | 200 | 5K | 0.97 | 1.85 | x | x | x | x | | 2 | C<br>DI | 256K | | x | 28 | B<br>D | x | | | |
| Sigma 5 | 350 | 8.5K | 0.95 | 3.1 | x | x | x | x | | 7 | C<br>DI | 512K<br>3b | 0.9<br>17ms | x | 256 | B | x | | | |
| Sigma 6 | 700 | 18K | 0.95 | 2.7 | x | x | x | x | x | 16 | C<br>DI | 512K<br>3b | 0.9<br>17ms | x | 256 | B | x | | | |
| Sigma 7 | 800 | 20K | 0.95 | 2.7 | x | x | x | x | x | 16 | C<br>DI | 512K<br>3b | 0.9<br>17ms | x | 256 | B | x | | | |

Source: *Infosystems,* 21, No. 5 (May 1974), pp. 56–57.

One vendor's line of computers is shown in Figure 9.6. It describes the many elements of a system that must be evaluated and compared. Figure 9.7 shows some of the characteristics that are evaluated when selecting minicomputers.

Figure 9.8 is a comparison chart of printing terminals manufactured by different companies. Figure 9.9 illustrates the types of characteristics of microfilm retrieval units that are compared in equipment selection.

A comparison of computer software is illustrated in Figure 9.10. It shows the features of the data base management systems that are available.

## EXERCISES

1. What elements are taken into consideration in defining a problem? Give several examples.
2. What is the function of a feasibility study?
3. What are the three major phases of the feasibility study?
4. List at least three different approaches that may be followed in assigning staff to conduct the feasibility study.
5. What is the function of the preliminary study?
6. What is the function of the investigative study?
7. What is the function of the final report?
8. What elements are considered when comparing computer software for a business system?
9. What elements are evaluated when considering computer hardware?
10. What are computer utility programs?
11. Conduct a short feasibility study of an activity such as purchasing a new or used car.
12. Visit a business firm that has conducted, or plans to conduct, a feasibility study. Discuss the aims of the study and how people were selected to conduct it.
13. Assume you are starting a small retail firm. Conduct a feasibility study of various methods of writing and processing orders.
14. Study your school's registration system. Prepare a feasibility study showing an improved method of registering students.
15. Conduct a feasibility study comparing installation of manual and electric typewriters for a small office.

**Figure 9.7.** Small business computer comparison chart

# 1973 SMALL BUSINESS COMPUTER SPECIFICATIONS CHART

• Standard
† Optional

| MANUFACTURER | Reader service number | Model number | Purchase price (in thousands of dollars) | Monthly lease/rental price | CAN BE USED FOR | | | | | | SOFTWARE PACKAGES AVAILABLE | | | | | | | | | | USER PROGRAMMING | | | | | | | SUPPLIER SERVICES | | | | INPUT/OUTPUT DEVICES | | | | | | | |
|---|---|---|---|---|---|---|---|---|---|---|---|---|---|---|---|---|---|---|---|---|---|---|---|---|---|---|---|---|---|---|---|---|---|---|---|---|---|---|---|
| | | | | | Single-entry processing | Batch processing | Communications | Remote inquiry | Multi-programming | Direct inquiry | Billing & invoicing | Inventory | Payroll | Order entry | Accounting/general ledger | Payables | Receivables | Sales analysis | Other | Industry packages available | COBOL | FORTRAN | RPG | BASIC | Assembler | Machine | Other | Classroom training | On-site training | System design service | Programming service | Keyboard/printer | Punched card | Magnetic card | Paper tape | Magnetic tape | Magnetic disk or drum | Line printer | CRT or other display |
| Basic/Four Corp. | 346 | 350 | 30.9 | 680 | •ff | • | • | • | • | • | • | • | • | • | • | • | • | • | | • | | | | • | | | | • | • | • | • | | • | | • | • | • | • | • |
| | | 400 | 31.9 | 702 | •ff | • | • | • | • | • | • | • | • | • | • | • | • | • | | • | | | | • | | | | • | • | • | • | | • | | • | • | • | • | • |
| | | 500 | 32.9 | 724 | • | • | • | • | • | • | • | • | • | • | • | • | • | • | | • | | | | • | | | | • | • | • | • | | • | | • | • | • | • | • |
| Burroughs | 393 | L8000 | 13-50 | 355-1000 | • | • | • | • | | • | • | • | • | • | • | • | • | • | • | nn | • | | | | | | | | • | • | • | • | • | | • | • | • | • | |
| | | B700 | 42-110 | 950-2600 | • | • | • | • | | • | • | • | • | • | • | • | • | • | • | nn | • | | | | | | | | • | • | • | • | • | | • | • | • | • | |
| | | B1712 | 70-120 | 1500-2800 | • | • | | | • | • | • | • | • | • | • | • | • | • | • | nn | • | • | • | • | • | | | • | • | • | • | • | • | | | • | • | • | |
| | | B1714 | 75-200 | 1600-3500 | • | • | • | • | • | • | • | • | • | • | • | • | • | • | • | nn | • | • | • | • | • | | | • | • | • | • | • | • | | | • | • | • | |
| | | B1726 | 135-475 | 3000-10,000 | • | • | • | • | • | • | • | • | • | • | • | • | • | • | • | nn | | | | | | | | • | • | • | • | • | • | | • | • | • | • | |
| Business Controls Corp. | 348 | System 80 | 40-100 | 800-2000 | • | • | • | • | • | • | • | • | • | • | • | • | • | • | • | • | | • | | • | • | | | | • | • | | • | • | | | • | • | • | • |
| Cascade Data, Inc. | 399 | Concept II 3010 | 27.5 | 550 | • | • | | | | • | • | • | • | • | • | • | • | • | • | • | | | • | | • | | | • | • | • | • | • | • | | • | • | | • | |
| | | Concept II 4010 | 38.3 | 766 | • | • | | | | • | • | • | • | • | • | • | • | • | • | • | | | • | | • | | | • | • | • | • | • | • | | • | • | • | • | • |
| | | Concept II 4020 | 52 | 1040 | • | • | | • | • | • | • | • | • | • | • | • | • | • | • | • | | | • | | • | | | • | • | • | • | • | • | | • | • | • | • | • |
| | | Concept II 4030 | 64.7 | 1295 | • | • | • | • | • | • | • | • | • | • | • | • | • | • | • | • | | | • | | • | | | • | • | • | • | • | • | | • | • | • | • | • |
| IME Checkwriter Co. | 405 | IME 10001 | 15-31 | d | • | | | | | • | • | • | • | • | • | • | • | • | • | | | | | | | • | | • | • | • | • | • | | • | | • | | | • |
| Clary Datacomp | 349 | Business System 404 | 30-40 | 1000k | • | • | • | • | | • | • | • | • | • | • | • | • | • | •j | | | | • | • | • | • | | • | • | • | • | • | | • | • | • | • | • | • |
| Custom Computer Systems, Inc | 351 | Simplex-70 Model 8 | 59.4 | 990-1165 | • | • | • | • | • | • | • | • | † | • | † | † | • | • | •r | | | • | | | • | | | • | • | • | • | • | • | | • | • | • | • | • |
| | | Simplex-70 Model 12 | 65.7 | 1090-1290 | • | • | | • | • | • | • | • | † | • | † | † | • | • | •r | | | | | | • | | | • | • | • | • | • | | | • | | • | • | • |
| Data General Corp. | 352 | 840 | 16.5h | N/A | • | • | • | • | • | • | | | | | | | | | | | | • | | • | • | • | •i | • | • | | | • | • | | • | • | • | • | • |
| | | Nova 820 | 6.1g | N/A | • | • | • | • | • | • | | | | | | | | | | | | • | | • | • | • | •i | • | • | | | • | • | | • | • | • | • | • |
| | | Nova 1220 | 4.9g | N/A | • | • | • | • | • | • | | | | | | | | | | | | • | | • | • | • | •i | • | • | | | • | • | | • | • | • | • | • |
| | | Nova 1210 | 4g | N/A | • | | • | | | | | | | | | | | | | | | • | | • | • | • | •i | • | • | | | • | • | | • | • | • | • | • |
| Datapoint Corp. | 400 | Datapoint 2200 | 6 | | • | • | • | • | | • | | | | | | | | | | | | | • | • | • | • | •xx | • | • | • | | • | • | | | • | • | • | • |
| Data Trends, Inc. | 353 | System A | 32.8 | | • | • | | | • | • | • | • | • | • | • | • | • | • | | • | | | | | | • | | • | • | • | • | • | | | | • | | • | • |
| | | General Terminal Units | 6.2 | | • | | • | | | • | • | • | • | • | • | • | • | • | | | | | | | | • | | • | • | | • | • | | | | • | | • | • |
| Digital Computer Controls Inc. | 354 | D-112 | 3-6 | N/A | | | • | | | | | | | | | | | | | | | | | • | • | • | | • | • | • | • | • | • | | • | • | • | • | • |
| | | D-116 | 3.2-7 | N/A | | | • | | | | | • | • | | • | | | • | | | | • | | • | • | • | | • | • | • | • | • | • | | • | • | • | • | • |
| | | D-116H | 3.6-8 | N/A | | | • | | | | | • | • | | • | | | • | | | | • | | • | • | • | | • | • | • | • | • | • | | • | • | • | • | • |
| | | D-112H | 3.3-6.8 | | | | • | | | | | | | | | | | | | | | | | • | • | • | | • | • | • | • | • | • | | • | • | • | • | • |
| Digital Equipment Corp. | 403 | Dec Data-system Series 300 | 30-87 | | • | • | | | | • | | | | | | | | | | | | † | | † | † | | •uu | • | • | • | • | • | • | | • | • | • | • | • |
| | | Dec Data-system Series 500 | 60-610 | | • | • | • | • | | • | | | | | | | | | | | | • | • | • | • | | | • | • | • | • | • | • | | † | • | • | • | • |
| Digital Scientific Corp. | 355 | Meta 4/1130 | 50.6 | 2197 | | • | | • | | • | | | | | | | | | | | • | • | • | | • | • | •gg | • | • | | | • | • | | • | • | • | • | † |
| Eldorado Computer Corp. | 401 | 140 | 21 | | • | • | • | • | • | • | • | • | • | • | • | • | • | • | • | • | | | | • | • | • | | • | • | • | • | • | • | | • | • | • | • | • |
| Four-Phase Systems | 358 | System IV/40 | 15.7s | 315s | • | • | • | • | | • | | | | • | | | | | •u | • | • | | | | • | • | • | • | • | • | • | • | • | | | • | • | • | • |
| | | System IV/70 | 27.9 | 620 | • | • | • | • | | • | | | | • | | | | | •u | • | • | | | | • | • | • | • | • | • | • | • | | | | | • | • | • |

**Figure 9.8.** Printing terminal comparison chart

## Feature comparison chart – printing terminals

| Feature name | Memorex 1280 | Novar 5-50 | Teletype 37/352-2X | G. E. Terminet 300 | Memorex 1240 |
|---|---|---|---|---|---|
| 1. Max. print speed | 120 char./sec. | 15 char./sec. | 15 char./sec. | 15 char./sec. | 120 char./sec. |
| 2. Mean maintenance interval | 45 days | 1 month | 6 months | 6 months | 45 days |
| 3. Max. transmission rate to tape | 120 char./sec. | 180 char./sec. | 120 char./sec. | 120 char./sec. | n.a. |
| 4. Max. print positions | 120 positions | 130 positions | 132 positions | 118 positions | 120 positions |
| 5. Available print characters | 94 graphics | 96 graphics | 94 graphics | 94 graphics | 48/94 graphics |
| 6. Tractor drive | yes | yes | yes | yes | yes |
| 7. Adjustable tractor drive | yes | yes | no | no | yes |
| 8. Maximum form parts | 6 parts | 6 parts | 6 parts | 6 parts | 6 parts |
| 9. Service availability | L.A., S.D. | S.D., L.A. | L.A. | L.A. | L.A., S.D. |
| 10. Purchase price with options | $10,335 | $9,640 | $5,570 | $4986 + $1300 = $6286 | $8,420 |
| 11. Price/print speed ratio | 86:1 | 642:1 | 371:1 | 419:1 | 70:1 |
| 12. Monthly lease rate (w. options) | $291 | $281 | not available | $182 | $196 |
| 13. Present number of installations | new-1st Inst. Sept. 71 | 2000 | 10,000 | 12,000 | 200 |
| 14. Tape cartridge capacity | 90,000 char. | 73,000 char. | 150,000 char. | 50,000 char. | n.a. |
| 15. Lead time after order | 6 months | 1 month | 4 months | 2 months | 1 month |
| 16. Remote tab set/clear | yes | no | yes | yes | yes |
| 17. Operator instruction | by manufacturer | by manufacturer | by manufacturer | by manufacturer | by manufacturer |
| 18. Transmission codes available | ASCII | EBCD, BCD | ASCII | ASCII | ASCII |
| 19. Data set (s) or equivalent | 202D or 103A | 202 or 103 | 103 & 202 | 202 or 103 | 202 or 103 |
| 20. Cost/mo. of data set | 1222-$20, 1220-$10 | included in rental | $70 | $70 | 1222-$20, 1220-$10 |
| 21. Variable transmission speed | 10, 15, 30 & 60 cps | 13.4, 30, 180 cps | fixed for tape & printer | 11, 15, & 30 cps | 10, 15, 30, & 60 cps |
| 22. Instruction set | full ASCII | ASCII available | full ASCII | full ASCII | full ASCII |
| 23. Polling and addressing avail. | yes | yes | yes | no | yes on 1251 |
| 24. Longitudinal redundancy check | yes | yes | no | no | no |
| 25. Cartridge cost | $5 ea. | $15 ea. | $8 ea. | $5 ea. | n.a. |
| 26. Printing method | chain printer | spherical printer | moving type set | chain printer | chain printer |
| 27. Buffer capacity | 256 characters | 350 characters | no buffer | no buffer | no buffer |
| 28. Cartridge type | Phillips | Digitronic | 1/2" tape-spool in/out | Philips | n.a. |
| 29. Simultaneous tape/print oper. | no | no | yes | no | n.a. |
| 30. Handle card stock forms | no | yes | yes | no | no |
| 31. Required number of ports | 2 ports w. 202 | 2 ports w. 202 | 3 ports w. 103 & 202 | 2 ports w. 202 | 2 ports w. 202 |
| 32. Vertical redundancy check | yes | yes | no | yes | yes |
| 33. Vertical forms control | yes | no | yes | yes (disc. control) | yes |
| 34. Compatible with NCR 621-101 | yes | not tested | yes | not tested | yes |

n.a. = not applicable

| CDC 712 | Sanders 3120 | Motorola MTP | 6000 | Kleinschmidt-311 | Memorex 1252 | Univac DCT 500 | |
|---|---|---|---|---|---|---|---|
| 30 char./sec. | 165 char./sec. | 240 char./sec. | | 27 char./sec. | 120 char./sec. | 30 char./sec. | 1. |
| 7 days | 30 days | | | 3-4 months | 45 days | 30 days | 2. |
| n.a. | n.a. | n.a. | | n.a. | n.a. | n.a. | 3. |
| 132 positions | 132 positions | 100 positions | | 72 positions | 120 positions | 132 positions | 4. |
| 64 graphics | 63 graphics | 94 graphics | | 64 graphics | 94 graphics | 63 graphics | 5. |
| yes | yes | no | | yes | yes | yes | 6. |
| yes | yes | no | | no | yes | yes | 7. |
| 6 parts | 6 parts | 1 part | | 3 parts | 6 parts | 6 parts | 8. |
| L.A., S.D. | S.D., L.A. | L.A. | | L.A. | L.A., S.D. | L.A., S.D. | 9. |
| $4,950 | $9,130 | | | $3,938 | $11,345 | $4,485 | 10. |
| 165:1 | 55:1 | | | 146:1 | 95:1 | 162:1 | 11. |
| $150 | $431 (inc. main.) | | | $282 | $261 | $115 | 12. |
| new-12 installations | new | | | 8,000 | new | 400-500 | 13. |
| n.a. | n.a. | | | n.a. | n.a. | n.a. | 14. |
| 4 months | 4-5 months | | | 4-6 months | 3 months | 3 months | 15. |
| line printer | yes | | | no | yes | no | 16. |
| by manufacturer | by manufacturer | | | by manufacturer | by manufacturer | by manufacturer | 17. |
| ASCII, BCD | ASCII | | | ASCII (18 level) | ASCII | ASCII | 18. |
| 201 | 201 | 202C2, 202D2 | | 103A | 202D or 103A | 103A2 or 103F | 19. |
| $70 | $70 | $30 | | $25 | 1222-$20, 1220-$10 | $15 | 20. |
| variable | 240, 200 to 30 cps | | | 40 or 24 cps | 10, 15, 30 & 60 cps | 10, 15, & 30 cps | 21. |
| ASCII | ASCII | ASCII | | ASCII | full ASCII | ASCII | 22. |
| yes | no | | | no | yes | yes | 23. |
| yes | yes | | | no | yes | no | 24. |
| n.a. | n.a. | | | n.a. | n.a. | n.a. | 25. |
| wheel printer | matrix | matrix (series) | | wheel printer | chain printer | helical wheel | 26. |
| 192 characters | 768 or 1024 char. | | | no buffer | 256 char. (expandable) | no buffer | 27. |
| n.a. | n.a. | | | n.a. | n.a. | n.a. | 28. |
| n.a. | n.a. | | | n.a. | n.a. | n.a. | 29. |
| yes | yes | | | no | no | yes | 30. |
| 2 ports w. 201 | 2 ports | | | 1 port | 2 ports w. 202 | 2 ports | 31. |
| yes | yes | | | no | yes | yes | 32. |
| yes (start of form only) | yes (2 channel tape) | | | yes (additional cost) | yes | no | 33. |
| not tested | not tested | | | not tested | not tested | not tested | 34. |

n.a. = not applicable

Source: *Computer Decisions, 5,* No. 4 (April 1973), pp. 24–25.

**Figure 9.9.** Microfilm retrieval unit comparison chart

• Standard

| MANUFACTURER | Model or series | OPERATES AS: Reader | Printer | Reader-printer | Price | WORKS WITH THESE MEDIA: 16 mm cartridge | 16 mm reel | 35 mm | 70 mm | 105 mm | Aperture cards | Jackets | Micro-opaques | Micro-positives | Other | FEATURES OF READERS AND PRINTERS — DATA LOCATING: Visual | Automatic or semi-automatic | Other | Dimensions of reader screen | COPY MAKING CAPABILITIES: Time to make and deliver copy (seconds) | Cost for one copy (cents) | Process for making copies |
|---|---|---|---|---|---|---|---|---|---|---|---|---|---|---|---|---|---|---|---|---|---|---|
| AM Bruning Div. | 150 | • | | | | | | | | | • | | | | | | | | 11x9 | | | |
| | 175 | • | | | | | | | | | • | | | | | | | | 6x9 | | | |
| | 200 | • | | | | | | | | | • | | | | | | | | 16x12 | | | |
| | 220 | • | | | | | | | | | • | | | | | | | | 16x12 | | | |
| | 1200 | | • | | | | | • | | | • | | | | | | • | | | b | | electrostatic |
| | 500 | • | | | | | | | | | • | | | | | | | | 14x20 | | | |
| | 95 | • | | | | • | | | | • | | | | | | | • | | 14x14 | | | electrostatic |
| Agfa-Gevaert, Inc. | Copex LP-3 | • | | | | | | | | | | • | | | | • | | | 16x13 | | | |
| Bell & Howell Co. | SR-I | • | | | c | | | | | • | • | • | | | | • | | | 9x11 | | | |
| | SR-II | • | | | c | | | | | • | • | • | | | | • | | | 11x9 | | | |
| | SR-III | • | | | c | | | | | • | • | • | | | | • | | | 19x11 | | | |
| | SR-IV | • | | | c | | | | | • | • | • | | | | • | | | 15x11 | | | |
| | SR-V | • | | | c | | | | | • | • | • | | | | • | | | 7x9 | | | |
| | Spacemaster | | | • | c | | | | | • | • | • | | | | • | | | 11x11 | 8 | 2–5 | electro-photo |
| | Autoload | | | • | c | • | • | | | | | | | | | • | | • | 11x11 | 8 | 2–5 | electro-photo |
| | Reporter | | | • | c | | • | | | • | • | • | | | | • | | • | 11x11 | 8 | 2–5 | electro-photo |
| | Autoload | • | | | c | • | • | | | | | | | | | • | | • | 14x14 | | | |
| | UV-40 | • | | | c | • | • | | | | | | | | | • | | • | 14x14 | | | |
| | Briefcase | • | | | c | | | | | • | • | • | | | | • | | • | 7x9 | | | |

| | | | | | | | | | | | | | | | | | | | |
|---|---|---|---|---|---|---|---|---|---|---|---|---|---|---|---|---|---|---|---|
| Canon U.S.A., Inc. | 330FT | | ● | | | ● | | | | | ● | | ● | | | 11x11 | 13 | | stabilization |
| | 330MB | | ● | | ● | | | | | | | | | ● | | 11x11 | 13 | | stabilization |
| A.B. Dick Co. | 810 | | ● | 2600 | | | ● | ● | ● | | ● | | ● | | | 12x16 | 10 | 2–4 | electrofax |
| Eugene Dietzgen Co. | 4305 | ● | | | | | | | | ● | | | | | k | 11x12 | | | |
| | 4309 | ● | | | | | | | | ● | ● | a | | | k | 11x9 | | | |
| | 4311 | ● | | | ● | ● | | | | | | | | | k | 16x16 | | | |
| | 4313-AR | ● | | | | ● | ● | | | ● | ● | a | | | k | 18x24 | | | |
| | 4316 | ● | | | | | | | | ● | ● | a | | | k | 11x12 | | | |
| | 4322 23 24 | ● | | | | | | | | ● | ● | a | | | k | x | | | |
| Dukane Corp. | 27A25B | ● d. | | 373 | | ● | ● | | | ● | ● | a | ● | | | 14x14 | | | |
| | 576-90 | ● e | | 200 | | | | | | ● | | | | | | | | | |
| Eastman Kodak Co. | Recordak | ● | | | | | ● | | | ● | ● | a | ● | | | f | | | |
| | Film Readers | | | | | | | | | | | | | | | | | | |
| | Recordak | ● | | | | ● | | | | | | | ● | | | 9x12 | | | |
| | Starmatic | | | | | | | | | | | | | | | | | | |
| | Recordak | ● | | | | | | | | ● | ● | a | ● | | | 11x16 | | | |
| | Easematic | | | | | | | | | | | | | | | | | | |
| | Recordak | | ● | | ● | ● | ● | | | ● | ● | a | | ● | | 11x16 | 28 | 10 | silver |
| | Magnaprint | | | | | | | | | | | | | | | | | | monobath |
| | Recordak | ● | ● | | ● | | | | | | | | ● | ● | | 14x14 | 10 | 3.2 | zinc oxide |
| | Microstar | | | | | | | | | | | | | | | | | | |
| | Recordak | ● | ● | | ● | ● | ● | | | ● | ● | a | ● | | | g | 12 | 3.2 | zinc oxide |
| | Motormatic | | | | | | | | | | | | | | | | | | |
| | Recordak | ● | | | | | | | | ● | | | ● | | | 11x12 | | | |
| | MKR-I | | | | | | | | | | | | | | | | | | |
| | Recordak | | ● | | | | | | | ● | ● | a | ● | | | 11x11 | 28 | 10 | silver |
| | Magnafiche | | | | | | | | | | | | | | | | | | monobath |
| | Miracode II | ● | ● | | ● | | | | | | | | | ● | | 14x14 | 10 | 3.2 | electrostatic |
| | Ektalite | ● | | | | | | | | ● | ● | a | ● | | | 6x8 | | | |
| | 120 140 | | | | | | | | | | | | | | | | | | |
| | Ektron | | ● | | | | ● | | | ● | | | ● | | | 18x24 | 30 | 14 | electrostatic |
| Ednalite Corp. | 1624 | ● | | 620 | ● | ● | | | | | | | | | h | 12x14 | | | |
| | 1625 | ● | | 770 | ● | ● | | | | | | | | | h | 12x14 | | | |
| | 1640 | ● | | 895 | ● | ● | | | | | | | | | h | 12x14 | | | |
| | 1625RP | | ● | 1.850 | ● | ● | | | | | | | | | h | 12x14 | 20i | 5j | electrostatic |

**Figure 9.9**—Continued

• Standard

| MANUFACTURER | Model or series | OPERATES AS: Reader | Printer | Reader-printer | Price | WORKS WITH THESE MEDIA: 16 mm cartridge | 16 mm reel | 35 mm | 70 mm | 105 mm | Aperture cards | Jackets | Micro-opaques | Micro-positives | FEATURES OF READERS AND PRINTERS — DATA LOCATING: Other | Visual | Automatic or semi-automatic | Other | Dimensions of reader screen | COPY MAKING CAPABILITIES: Time to make and deliver copy (seconds) | Cost for one copy (cents) | Process for making copies |
|---|---|---|---|---|---|---|---|---|---|---|---|---|---|---|---|---|---|---|---|---|---|---|
| GAF Corp. | 9000 | | | • | 995 | | | | | • | • | • | | | | • | | | 11x11 | 8 | | electrostatic |
| | COR701 | • | | | 913 | • | | | | • | • | • | | | | • | • | | 17x12 | | | |
| | 7500 | • | | | 129–265 | | | | | • | • | • | | | | • | | | | | | |
| | 7600 | • | | | 260–900 | | | | | | • | | | | | • | | | | | | |
| Karl Heitz, Inc. | Optico L24 | • | | | 399 | | • | • | | | | | | | | • | | k | 9x9 | | | |
| | Optico L26 | • | | | 699 | | • | • | | | • | • | | | | • | | k | 14x14 | | | |
| Image Systems, Inc. | 201 | • | • | • | | | | | | • | | | | | a | | • | | 14 | 10 | | electrostatic |
| Information Design, Inc. | 201-1 | • | | | 1080 | • | • | • | | | | | | | I | • | | | 24x24 | | | |
| | 201-M | • | | | 1280 | • | • | • | | | | | | | I | • | | | 24x24 | | | |

Reprinted from the May 1974 issue of *Modern Office Procedures* and copyrighted 1973 by Industrial Publishing Company, Division Pittway Corporation.

**Figure 9.10.** Data base management system comparison chart

# Data base management systems

| Company | System | Cost Lease/purchase | Systems supported | Core required | File organization | User language | System language | Number of installations | Date first installed | Circle No. |
|---|---|---|---|---|---|---|---|---|---|---|
| **Cincom Systems, Inc.** 2181 Victory Pkwy. Cincinnati, OH 45206 (513) 961-4110 | Total | $750 / $24,500 | 360, 370 Series 70 H1200 | 14k | Random | Any (with call or exit functions) | BAL | 200+ | 1968 | 301 |
| **Computer Corp. of America** 575 Technology Square Cambridge, MA 02139 (617) 491-3670 | IFAM | $200 / $10,000 | 360, 370 | 48k | Inverted Index sequential Random sequential | Fortran Cobol PL/1 BAL | BAL | 3 | 1972 | 302 |
| | 204 (on-line) | $300 / $15,000 | 360, 370 | 80k | Inverted Index sequential Random sequential | Own | BAL | 3 | 1968 | 303 |
| **Cybertech Data Systems,** 3000 Diamond Park Dallas, TX 75247 (214) 638-6580 | Re-act | N.A. / $15,000 | 360, 370 | 22k | Sequential Index sequential | Own | BAL | 22 | 1970 | 304 |
| **Infodata Systems, Inc.** 1901 Ft. Myer Dr. Arlington, VA 22209 (703) 524-6700 | Inquire | $1100 / $28,500 | 360, 370 | 110k | Direct access Index sequential | Own | PL/1 | 28 | 1969 | 305 |
| **Informatics, Inc.** 21050 Vanowen St. Canoga Pk., CA 91303 (213) 887-9121 | Mark IV | N.A. / $35,000 | 360, 370 Series 70 | 32k | Sequential Index sequential Direct access | Own | BAL | 500 | 1967 | 306 |
| **International Business Machines Corp.*** 1133 Westchester Ave. White Plains, NY 10604 (914) 696-1900 | IMS/360 | $550 / N.A. | 360, 370 | 80k | Sequential Index sequential Direct access | Cobol PL/1 BAL | BAL | 400* | 1970 | 307 |
| **Mathematica, Inc.** Box 2392 Princeton, NJ 08540 (609) 799-2600 | Ramis | $840 / $21,000 | 360, 370 | 132k | Special file structure | Fortran Cobol Assembler PL/1 | BAL and Fortran | 70 | 1967 | 308 |
| **McDonnell Douglas Automation Co.** 309 Evergreen Building Renton, WA 98055 (206) 228-5353 | GSS | $540 / $18,000 | 360, 370 | 54k | Sequential Index sequential | Own Assembler Cobol | BAL | 22 | 1969 | 309 |
| **MRI Systems Corp.** Box 9510 Austin, TX 78766 (512) 258-5171 | System 2000 | $980 / $25,000 | 360, 370 1106, 1108 CDC 6000 Cyber 70 | 130k | Special file structure | Assembler Cobol Fortran | Fortran and BAL | 50 | 1970 | 310 |
| **Programming Methods, Inc.** 1301 Avenue of the Americas New York, NY 10019 (212) 489-7200 | Score III | N.A. / $12,500 | 360, 370 B2500/3500 Series 70 1108/9000 NCR Century CDC 3000 | 32k | Sequential Index, sequential | Cobol Own | Cobol | 300+ | 1967 | 311 |
| **Software AG** 12124 Basset Lane Reston, VA 22070 (703) 471-5098 | ADABAS | $4,500 / $120,000 | 360, 370 Series 70 | 110k | Any | Own Cobol Fortran PL/1 BAL | BAL | 21 | 1971 | 312 |
| **System Development Corp.** 2500 Colorado Ave. Santa Monica, CA 90406 (213) 393-4411 | DS/2 | $475 / $15,000 | 360, 370 | 30k | Sequential Index sequential | Own | BAL | 17 | 1971 | 313 |
| **United Aircraft Res. Labs** Silver Lane East Hartford, CN 06108 (203) 565-5447 | UAIMS | $800 / $24,000 | 360, 370 | 60k | Direct access Inverted | Fortran Cobol BAL | Fortran and BAL | 4 | 1970 | 314 |

* Other main-frame manufacturers, such as Burroughs, Control Data, Honeywell, Univac and Xerox, offer similar systems or their own computers.

* estimated

Source: *Computer Decisions*, 4, No. 8 (August 1972), p. 15.

# Chapter 10

# Implementing the System

The goal of system implementation is to transform the plans, schedules, and designs from the feasibility study into an integrated, functioning operation. Many diverse and complex factors are involved. Psychological reactions must be anticipated and handled. Time schedules must be adhered to or adjusted. Training programs must be planned and instituted. Interruptions in office procedures and services must be minimized. And a host of other expected—and unexpected—problems must be managed.

The systems designer should be fully aware that even a small change in a system may have dramatic effects upon other parts of the system. More than one system failure has been traced to poor implementation, rather than to weaknesses in the system itself.

## THE TIME FRAME

One of the first considerations is the time frame selected for system implementation. In planning this time schedule, the designer should establish certain goals, or benchmarks, to serve as checkpoints during implementation. Setting specific dates for the completion of each phase is an important precaution against time losses and resulting financial complications.

Three basic patterns are employed when installing a new system: an immediate, all-at-once changeover; a gradual, step-by-step changeover; or a period of concurrency, when the new and old systems operate at the same time. These three approaches and their advantages and limitations are described below.

### All-At-Once Changeover

In this approach (Figure 10.1A) the old system is abandoned, and the new system becomes completely operational on a given date. All planning and design, purchasing, training, ordering, and the like are finished before this date, so that the new system is ready to go at the moment the old one ceases to operate.

The major advantage of this approach is cost. Since one system is dropped as the other is started, the firm pays for operating only one system at a time. No backup or temporary system costs are involved. The all-at-once changeover also occurs rapidly. The adjustment period to the new system may be difficult and trying, but once made it is complete and the period of disruption is kept to a minimum.

The all-at-once changeover is often completed in less time than is required by the other methods. The benefits of the new system can be realized at once. Having an improved system operational before competition can institute a similar system is often a clear advantage in a competitive marketplace.

A major limitation in this approach is the suddenness and abruptness of the change itself, which may not give employees sufficient time to adjust.

**Figure 10.1.** Time frame for system implementation

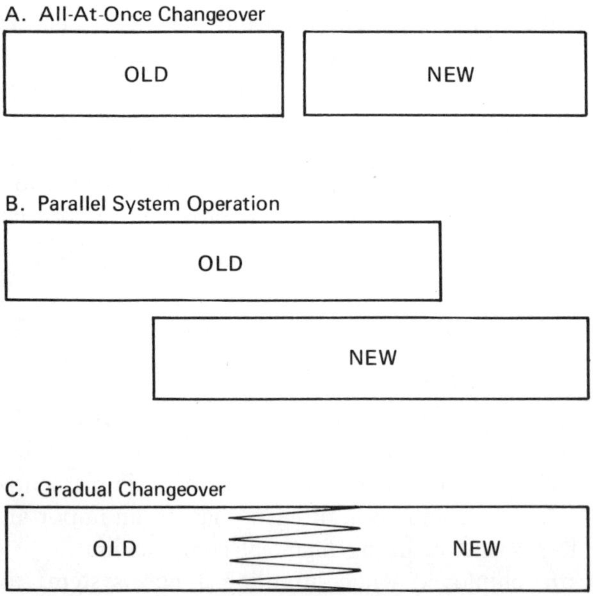

A. All-At-Once Changeover

OLD NEW

B. Parallel System Operation

OLD

NEW

C. Gradual Changeover

OLD NEW

Unexpected problems can also develop quickly, and with no other system to fall back on, can cause serious disruptions and confusion. Loss of data, errors in processing, goods misshipped, or interruption of service may present severe difficulties to customers, employees, and management, if no backup system is present.

### Parallel System Operation

Figure 10.1B illustrates a second approach to timing system implementation. In this method, both the old and new systems are operated concurrently for a period. The old system runs until the new one has been proven. Data is processed or moved through both systems concurrently. Only when the new system is fully checked out and operational is the old system abandoned.

Advantages to this approach are that the old system is available as a backup in the event of the new system's failure, and the results processed by the new system can be compared to the output of the old. In addition, changes and adjustments can be made in the new system without disturbing customer relations or order flow.

This approach is used in cases where loss of data or system failure cannot be tolerated. In such instances, the parallel system operation is continued until there is absolute certainty that the new one is functioning properly.

The major limitation is the cost involved in operating both systems during the transition period. The old system must be maintained and operated even after it is no longer productive. Extra shifts, overtime, and leasing duplicated equipment may be required.

Confusion is sometimes generated when two systems are operating simultaneously. Employees may not be sure which system to use, or which to trust. Orders may be misplaced, errors introduced, paperwork incorrectly duplicated. In some instances, operating parallel systems is impossible.

### Gradual Changeover

Figure 10.1C illustrates the gradual changeover. Instead of an abrupt change, the switch is made step-by-step. As one part of the new system is perfected and tested, that portion of the old system is shifted over. Then the next step is finished and that shift made. This process continues until the new system is fully implemented. For example, accounts payable may be put on the new system first, then receivables, payroll, inventory, and finally the last remaining functions.

A gradual changeover has many advantages. It gives the employees time to adjust to the new system. It allows personnel to learn details and ask questions without being rushed or being under the pressure of a chaotic change. A gradual changeover reduces the chances of a sudden, total systems failure. Since only part of the system is being implemented at any one time, no single catastrophic failure is likely to occur.

There are limitations to this approach, however. The gradual changeover may take a longer period of time—months, sometimes even years. In the meantime, the firm cannot realize the benefits or economic gains that should result from the new system. This may give a competitor, who has made an all-at-once change to an improved system, an advantage in the marketplace.

For example, the change to a new system may result in better service to the customer, faster shipment of goods, and fewer misplaced or misprocessed orders. While one company is making a gradual change to the new system, another makes an all-at-once change, gaining immediate benefits in terms of costs and improved customer relations.

Another limitation is that the full effects of the change may not be obvious for months. Thus, a malfunctioning condition may go unrecognized until the situation has become critical. For example, goods may be misshipped or misbilled for several months before the error is detected, and then it may be too late to undo the damage. An abrupt changeover would have shown up the error immediately, with a minimum of loss.

## PSYCHOLOGICAL CONSIDERATIONS

After a new system has been sold to management, the next step is to sell it to the people who will use it. As soon as word gets around that changes are brewing, psychological reactions, such as confusion, annoyance, apprehension, resistance, and even fear, can be expected. "What's wrong with the way we do things around here now?" probably expresses the common attitude of the grapevine. (See Figure 10.2.)

The systems analyst should understand why people react this way, and how these feelings can be changed to ones of cooperation, interest, and involvement. Many employees may feel threatened with loss of status, position, or power within the organization. When a change in procedure or structure occurs, others fear they will suffer loss of wages, or end up with degraded working conditions, increased inconveniences, longer hours, or exposure to criticism. Some fear a new system just because it is new. It means changing old ways and facing the unknown.

These reactions can lead to serious problems that can delay system implementation, hinder retraining programs, add considerably to costs and time elements, and even prevent the new system from being installed. One way to avoid such problems is to convince all involved individuals in advance that the new system will be to their benefit and will improve their working situation and relationships.

A fundamental rule of systems design is: talk to people early in the design stages and use this input to assist in system development. Personnel involved in the genesis of a new system are more apt to support the system and to aid its implementation.

**Figure 10.2.** Word travels fast among employees

A sound approach for encouraging positive personnel reactions when implementing a new system is:

1. ANNOUNCE THE NEW SYSTEM FORMALLY. Officially inform employees of the possibility of system changes. Do not let the news spread by itself through the grapevine. Explain how changes will affect the employees

and their working environment. This avoids rumors and confusion. (See Figure 10.3.)

2. Involve personnel early. Talk to people—secretaries, clerks, salesmen, foremen, customers—early in the planning stages. Find out their needs and feelings toward the new system.

3. Stress benefits to be gained from the new system. Describe, in terms the employees can understand, the advantages and benefits that should result when the new system is in operation: improved customer relations, increased sales volume, reduced errors. Explain any expected improvements in working conditions and relationships.

4. Periodically report progress. A newsletter or bulletin should be disseminated regularly to update employees on the present status of system implementation. This is particularly important if the new system involves many steps or a prolonged period of implementation.

5. Encourage positive support. Reward individuals who give cooperation and extra service during system implementation. Acknowledge

**Figure 10.3.** General Telephone and Electronic's information communication center

those who help by recognition of service and, in some instances, monetary rewards.

## TRAINING AND IN-SERVICE EDUCATION

Implementing a new system usually involves some form of education or retraining to prepare personnel to operate the new machines and to process the new procedures and routines. Many weeks or months may be spent in developing and then conducting this training program. The systems designer considers several elements when planning a training program:

1. WHICH ACTIVITIES IN THE NEW SYSTEM WILL REQUIRE SPECIAL TRAINING? The systems analyst must select the tasks or operations in the system for which the personnel may need special training. What new machines will be installed, and how many operators will be needed? What kind of training will be necessary to operate them? Will an intensive training course be needed, or just a demonstration? What new forms will be used, and which personnel should be instructed in their routing and use? Which employees will need explanations about new procedures and how to handle exceptions, failures, overloads, or maintenance?

2. DESIGN TRAINING CURRICULUM. The content and schedule of the training program must be spelled out in detail. This process may require several months and considerable research. The knowledge, skills, and attitudes that must be imparted to the personnel for each individual task must be defined. Since many activities will not be too different from those already being performed, many jobs in the new system will only require an orientation or demonstration. Others, involving operating new devices or carrying out unfamiliar procedures, may need intensive classroom study and a period of practice. Each step and objective in the training process must be expressed in clear, concise behavioral terms that can be tested and measured.

There are several optional ways to conduct this training. Some vendors and suppliers provide free, or low cost, training programs using their equipment. They offer classes in equipment operation, systems, or programming. Other vendors offer programmed instruction courses by mail. These courses are self-taught, and may be used by the trainee at home.

The systems department may design part, or all, of the training program. Instructors may have to visit equipment installations or vendors, write training manuals, duplicate instructional materials, and prepare teaching aids and visuals.

After the contents of the training program have been designed, many practical considerations must be solved. Instructors must be selected and trained. The site where the training programs will be held must be set up. Time schedules must be arranged. Important questions must be answered.

Should the classes be held during the work day, during lunch time, weekends, at night? Who should pay? If vendor training programs or programmed instruction courses are utilized, who should pay for them?

3. Select personnel to fill new positions. The method of selecting the individuals for the various positions and roles of the new system must be determined and implemented. Should members of the steno pool or the duplicating department be trained to operate the new typewriters? Or should new people who already have these skills be hired instead? What basis should be used for filling positions—appointment by department chairman, seniority, competitive examinations, previous training, or requests of employees themselves?

Several factors will have to be considered when determining how these decisions should be made. These include the company's personnel policies, psychological effects on employees and their morale, and the availability of trained personnel.

4. Proficiency and the learning curve. The director and systems analyst should be aware of the implications of the learning curve when implementing the training program. The learning curve illustrates the relationship between continued practice at a given task and the resultant gains in proficiency, speed, and accuracy.

Figure 10.4 illustrates a learning curve showing that the performance of an employee beginning a new task is relatively slow, with a high error rate. As the employee practices the task, speed and accuracy increase—up to a

**Figure 10.4.** Learning curve

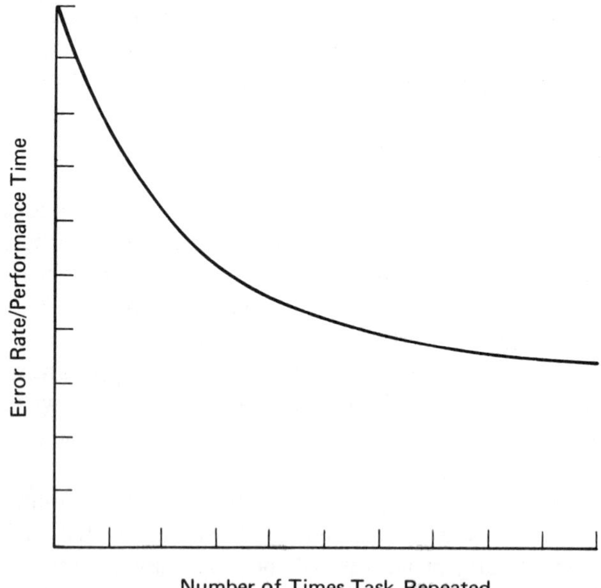

certain point, where a plateau is reached. Further practice will probably be of little use beyond this point.

As the learning curve indicates, performance is slow, the error rate is high, and output undependable when a new system is first installed. As employees learn to use the forms, equipment, and procedures, their speed and accuracy go up. After a certain period of time, however, performance stabilizes, and further practice yields little change in output.

It takes time to make a new system function. The system designer must expect delays, errors, questions, and problems when a new system is first installed. As the personnel gain facility in the new operations, these problems should subside. If they do not, they may be symptoms of other problems elsewhere in the system. A review of the steps involved in system design and implementation may be indicated.

## PRACTICAL CONSIDERATIONS

Unexpected, as well as expected, problems develop frequently during system implementation. While they can appear most anywhere, the systems analyst usually finds that certain areas are more generally the trouble spots. These include:

1. EXCESSIVE COSTS. Costs may increase much faster than anticipated during periods of change. Because of unfamiliarity with new routines, new sources of supply, new equipment, or lack of adequate accounting or inventory procedures, costs can sometimes be difficult to predict or control.

The systems analyst must be alert to this potential trouble spot. To avoid or minimize the problem, careful cost control and accounting measures must be instituted early. Costs from vendors for goods, services, and supplies must be watched closely during these periods of change.

2. INCONVENIENCE TO EMPLOYEES. A new system may bring many profitable improvements to a firm and its customers, but the short-term effects often bring inconvenience and disrupted working conditions. Implementation of a new system may require such interruptions in the normal flow of activity as overtime and extra hours, or rescheduled or cancelled vacations.

These problems should be anticipated and dealt with intelligently. For example, if a new system is to be implemented during the period when employees normally take vacations, the expected disruption should be announced well in advance. Employees need time to make changes or to revise their personal affairs.

If employees are expected to put in longer hours or to help during implementation, their roles should be spelled out clearly. Last minute recruitments or hastily drawn work schedules and assignments are the source of many problems, and lead to resistance and lack of cooperation.

3. INTERRUPTION OF SERVICE. Sometimes a new system causes temporary interruption of service or delivery to customers. In these instances, the customers should be forewarned of any possible inconvenience. Measures should be taken to reduce any disruption during the changeover. If necessary, alternate means of continuing services, such as contracting with outside vendors, should be considered. After the system has been implemented, the customers should be thanked for their cooperation and patience.

4. LEAD TIME. A most important aspect of systems implementation is lead time, the period between when a system or equipment is ordered and when it is delivered and installed. Neglecting to account for lead time when planning the new schedules can lead to serious disruption of service or delay in implementation.

Lead times vary with the type of service or equipment being purchased. The periods may range from one month to over a year for equipment. Some office machines or computer hardware, custom-made for a particular order, may involve months of design, engineering, and testing before final installation.

Computer programs used in a system may take months to write and test. A training program for employees may have a lead time of several months before it is ready for use. For instance, instructors must be trained, teaching materials prepared, and classroom facilities constructed.

Purchasing computers may mean an additional delay of several months, because benchmark programs must first be developed and tested. These test programs are run on several different vendors' equipment as a means of comparing computer speed and performance. Then, after the equipment has been selected and ordered, another period of time elapses before delivery.

For example, suppose a new communications system is to be installed. Some existing equipment will be utilized and some new equipment purchased. Figure 10.5 is a lead time schedule for such action. It shows the relative lead times required for each element in the system: specifying, ordering and installing equipment, testing, conversion, and changeover.

The first step is to prepare a project plan. Then several simultaneous activities begin. Equipment requisitions are issued, installation and maintenance plans formulated, test schedules set up, trouble reporting procedures planned, and the site preparation begun. Operator's manuals must be prepared and operators trained. Finally, testing and conversion procedures begin.

A review of this schedule shows how each step in the implementation is dependent on the other steps. If one element, such as site preparation, is not completed as scheduled, the entire project is delayed.

5. INSTALLATION AND TESTING OF HARDWARE. Another delay is often encountered between the time equipment arrives and the time it is placed in service. Office machines, computers, communications equipment, and

**Figure 10.5.** Lead time schedule

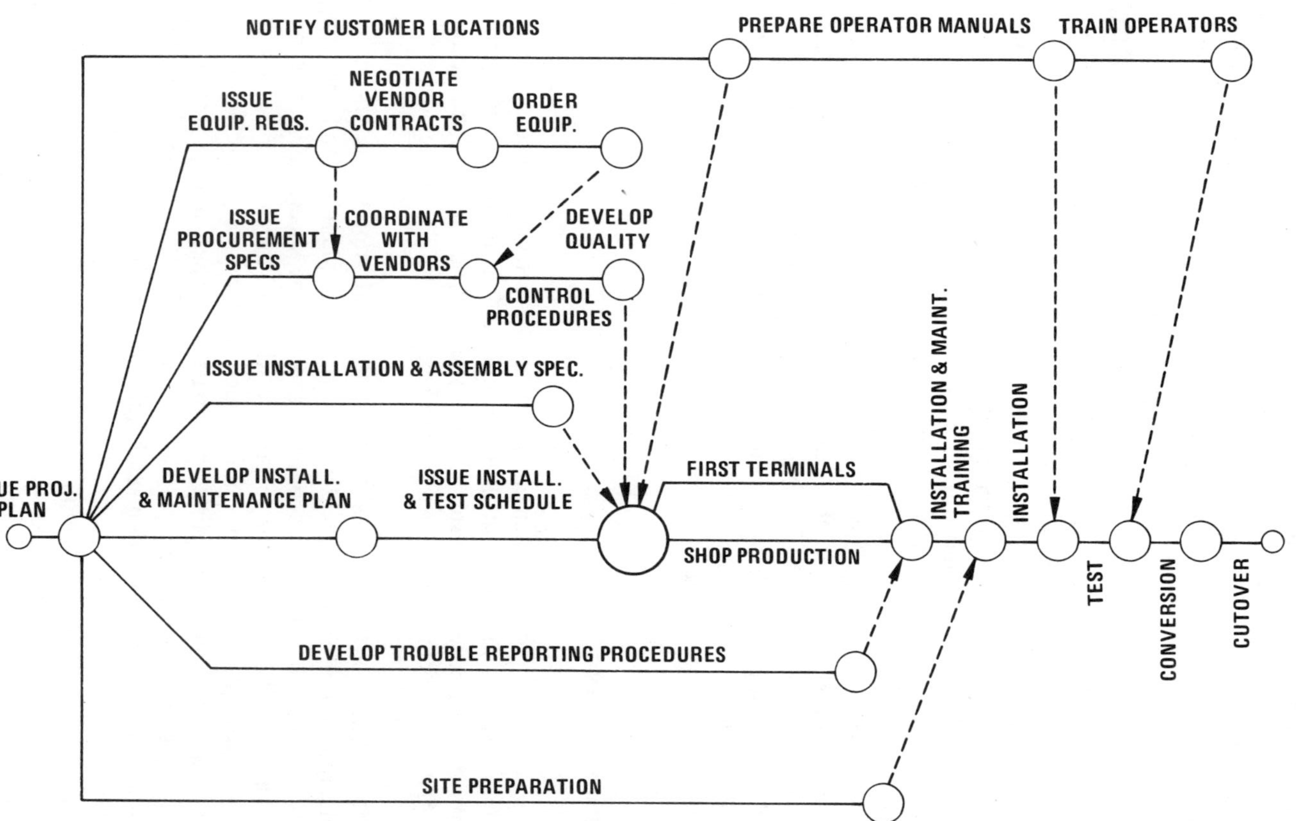


Source: *The Data Communications User, March* 1973, p. 26.

such must be thoroughly tested before they are fully operational. This may take several days or several months.

Lead time may be required to install special floors, air conditioning equipment, or wiring for a new computer system. After installation, the equipment may have to be tested on site. If the hardware does not meet expectations, it may have to be removed, redesigned, modified, and re-installed.

6. INSTALLATION AND TESTING OF SOFTWARE. Time must be allowed to design and order the software in a new system—computer programs, order processing routines, new forms. It may take many months of careful effort to write and "debug" a computer program. (Debugging—finding and removing errors from a program—can be very time consuming.) Test data must be prepared and run through the program, and the results compared to those computed manually from the same data or by the old system.

Any forms or records involved in the new system must be designed and set up. Proofs have to be checked for accuracy and revised if necessary. Often only a small quantity of the finished forms or records will be ordered at first for test use. Revisions and modifications in layout of entries or fields, or in physical movements from one work station to another, may be required. This process may take many weeks to complete.

## HELP DURING IMPLEMENTATION

The period of system implementation may be a trying one for customers, employees, and management. For many firms it is a unique and often disruptive experience. The systems analyst should be aware of the resources available for help during system implementation. These include:

1. HELP FROM VENDORS. Equipment and supply vendors are valuable resources for advice and assistance when planning and installing a new system. They have an economic stake in the success of a new system and are usually most willing to provide what aid they can.

For example, equipment vendors such as computer manufacturers may offer their technical expertise and assistance when a new installation is being planned. They help to design the computer system, monitor its installation, test its operation, and train the operators. Some provide fully tested and debugged programs along with the hardware.

Vendors run time and motion studies, cost analyses, and comparisons between different methods of data processing. Some make this information available to their customers at little or no charge. Vendors may prepare documents, reports, specifications, and suggested office layouts as resource material during system design.

Some vendors offer training programs to teach operators how to run their equipment. Classes in equipment operation, maintenance, and programming are held either at their site or at the customer's base.

In some instances, vendors make their own equipment, computers, or data processing machines available to the customer until the ordered equipment is delivered. This allows a firm to develop a system, such as a new billing or invoicing routine, and test and debug it on the vendor's equipment during the period that theirs is on order. When their equipment finally arrives, little time and effort is lost getting underway.

In addition, some vendors continue to make their own equipment available in case the customer suffers a system failure after installation. If a machine must be taken out of service for repairs or modification, the vendor provides backup facilities.

The systems analyst should remember that vendors are in business to make a profit, and successfully installed and operating systems are their best form of advertising.

2. HELP FROM CONSULTANTS. Another approach to systems implementation is to obtain the services of an outside consultant. In many locales, systems designers and implementation consultants offer their services to companies for a fee. They do not sell equipment, only services, and hence are in a more objective position to evaluate system hardware.

In many instances, it is more economical for a firm to hire a knowledgeable and experienced consultant to facilitate system implementation than to have its own personnel experiment and spend time and effort gaining the necessary knowledge.

3. STAFF RESOURCES. An often overlooked resource in implementing a new system is the pool of personnel already on the payroll. It is not uncommon for a company to turn to outside vendors, or a consultant, for assistance, only to overlook its own skilled and capable employees. Such a mistake can damage employee morale and endanger the implementation of the new system. Qualified employees involved in installing the system gain valuable information and experience useful for monitoring and maintaining systems during their regular operation.

The opposite situation, of course, may also be true. The system designer who relies solely on his or her own staff, regardless of capabilities and range of experience, may be making a serious mistake. It takes time to acquire the necessary information to do an adequate job of system implementation. Relying on unqualified or inexperienced personnel during a difficult period may delay implementation, add considerably to the cost, and produce an inferior system as the final result.

## EXERCISES

1. Describe the "all-at-once changeover."
2. Describe the parallel system changeover.
3. Describe the gradual changeover.
4. What psychological considerations must be evaluated when making a system change?
5. List three training considerations that must be evaluated when implementing a new system.
6. Describe some of the inconveniences that employees may face during a system change.
7. Define lead time and describe its effect upon system implementation.
8. List several sources of help that can be solicited during system implementation.
9. Describe the lead time involved in installing and testing new hardware.
10. Describe the lead time involved in installing and testing new software.
11. Suppose you were going to install a new data processing system in a business organization. Write a one-page statement, which will be distributed to employees, soliciting their cooperation and describing the new system.
12. Contact an equipment supplier and ask what training aids, manuals, and instructional materials are available with the equipment.
13. Discuss the effect of the learning curve on employee performance.
14. Visit an office machine store in your area and determine the lead times required to deliver various pieces of office equipment.
15. Assign another student a simple clerical task requiring about three minutes to execute. Ask him or her to carry it out several times. Time the activity and test the learning curve.

# Chapter 11

# Evaluating the System

Because an employee casually remarks that "Things have sure improved around here," doesn't necessarily mean that the analyst can then assume that the new system is operating properly and all problems have been solved. Without a careful evaluation of system output and achievements, and a comparison against the predetermined goals, no accurate or reliable conclusions about a system can be drawn at all.

System evaluation is the next step in system design, and it is an important and integral element. Its purpose is to measure the performance and output of a system quantitatively. These results are compared to the system goals established in the feasibility study. If the goals were met, system implementation has been successful. If they were not reached, the systems analyst must consider returning to an earlier phase in the process to reanalyze, redesign, or modify the elements in the system.

System evaluation is a frequently overlooked aspect of system design. The analyst will often expend great amounts of energy, effort, and skill to design and implement an elaborate system, and then assume that it is performing as expected. Without tools to measure with, and goals to measure against, the actual performance, improvements, and benefits achieved by a new system are only assumptions and hypotheses—not documented reliable facts.

## PERFORMANCE CRITERIA

An old expression says, "If you cannot measure it, you cannot control it." This is true in most areas of human endeavor, and in systems analysis as well. The first step in system evaluation is to establish precise performance criteria—if they have not already been defined. As a rule, these goals were established during the systems analysis stage before the system was implemented.

These goals must be stated in measurable terms. Statements such as "The new system is operating okay"; or "It offers better service to customers"; or "There are fewer troubles in the office"; are inadequate, and they offer no means of measurement or comparison. The degree of success or failure must be expressed in such terms as percentage of increase in productivity, improvement in the error rate, increase in the number of units produced in a given period, and increase in the number of orders processed by a given shift.

The criteria evaluated include the output and performance of the system as a whole, and of the individual elements that compose it. The results of the evaluation are compared to the estimated goals set earlier.

A system should operate through several complete cycles before it is evaluated. If a system is repeated several times daily, it can be evaluated after a few days. One that has a monthly cycle will not show reliable results until after several months have elapsed. Systems with an annual basis will require two to three years for an accurate assessment.

These are some of the common criteria used to measure the performance of a system:

### Time Element

The clock is an important tool of system measurement. It is a means of documenting the time units required for a particular action to be performed. It does not, however, measure the quality of the performance. For example, the clock can be used to measure the speed at which data moves through a particular work station or the time required for a computer program to be executed.

The clock may be applied to several aspects of system performance, including:

1. LEAD TIME. Sometimes called response time, lead time, as discussed earlier, is the time that elapses before a system responds to a demand placed upon it. In some systems, several hours or days may elapse before an order is acknowledged and its processing begun. Other systems may have virtually no delay before processing an order.

2. TURNAROUND TIME. Turnaround time is the period during which a system carries out a demand placed upon it. In other words, it is the length of time required before results are returned. A slow turnaround means a

comparatively long period of time is required for processing, and a fast turnaround means a short period is needed. Table 11.1 is a report showing that in this implementation, the actual turnaround time for small jobs was greater than estimated, but was less than estimated for longer jobs.

**Table 11.1**

| *Size of Job* | *Estimated Time* | *Actual Time* |
|---|---|---|
| Less than 10 pages or 10 minutes | 2 hours | 2 hours, 10 minutes |
| 11–30 pages or 11–30 minutes | 3 hours, 15 minutes | 3 hours, 32 minutes |
| 31–60 pages or 31–60 minutes | 6 hours | 4 hours, 8 minutes |
| More than 60 pages or 60 minutes | 24 hours | 18 hours, 30 minutes |

Lead time and turnaround time differ—a system can have a long lead time and a fast turnaround time—once the action is begun, results are returned promptly. Both time factors must be considered when evaluating system performance. Sometimes an increase in one is made at the expense of the other.

### Cost Element

The money involved in a system is used as a quantitative measure of performance—too often the only measure applied. Dollar costs are used to measure such things as profit, return on investment, errors in manufacture and shipping. Certainly a good system requires fewer dollars to service a given demand than does a poor system.

Costs are used to determine whether various parts of the system are performing up to financial expectations. Some common cost elements are:

1. LABOR. The amount of money that must be paid to labor for performing a given operation may be measured in dollars. Salaries, pay rates, and commissions are convenient quantitative measures of the human labor consumed in a system.

2. FACILITIES. System costs that do not change with the volume of output produced are called fixed costs, overhead, or burden. Facilities costs include expenses for machines, equipment, physical plant, office machines, and lighting. A system using considerable automatic equipment may have a low labor cost, but a high facilities cost. Another system, with many manual

operations or inadequate equipment, may have a high labor cost and a low materials cost. The cost of the facilities must be assessed and evaluated in relation to offsetting factors. Table 11.2 is a comparison of the facilities costs between an old and a new system.

**Table 11.2.** Monthly facilities costs analysis report

| *Description* | *Old System* | *New System* |
|---|---|---|
| Square footage | $1,900.00 | $1,200.00 |
| Utilities | 18.50 | 11.90 |
| Supplies | 360.00 | 510.00 |
| Equipment Lease | 300.00 | 670.00 |
| Air conditioning | 0.00 | 65.00 |
| Total — | $2,578.50 | $2,391.90 |
| Number of Jobs Processed — | 1,400 | 2,675 |

3. Materials. The costs in a system that vary with the volume of output are called materials, or variable, costs. They are incurred as goods are produced or as data is processed. Examples are the cost of paper, files, ribbons, and other supplies used during system operation. Variable costs differ from system to system. A computerized system might have a high facilities cost and a low materials cost.

4. Maintenance. After a system has been installed, the cost of maintaining the physical plant, office equipment, communications lines, and computer hardware and software must be considered. These costs will exist throughout the life of the system and must be examined during system evaluation. Table 11.3 is a report for a system where the actual emergency maintenance costs are less than estimated and the regular costs were accurately judged.

**Table 11.3.** Maintenance report

| | | *New System* | |
|---|---|---|---|
| | *Old System* | *Estimated Costs* | *Actual Costs* |
| Emergency Repairs | $3,100 | $550 | $450 |
| Regular Repairs | $110/month | $150/month | $150/month |

A successful system, of course, has a relatively low maintenance cost. Such costs must be balanced against labor and facilities costs during the system design and evaluation processes.

5. Expansion and modification. Costs are incurred when a system is modified, revised, or expanded. Some systems are easily expanded at a

low cost, while others may require extensive changes in design and equipment.

6. TRAINING. The costs of training operators and personnel to run a system must be assessed during system evaluation. Generally, the more complex the equipment in a system, the greater the training costs. Those considerably above expectations may suggest the need for reevaluating educational programs or personnel selection techniques.

7. DATA ENTRY. The costs of capturing data and entering it into the system must be measured in relation to other factors, such as output, processing, labor, equipment, and storage costs. Manual systems may involve sizeable data entry costs and high labor costs, but may have low equipment investment. Automated data entry systems, such as those using OCR or MICR, may involve low data entry costs, but have higher equipment expenses and may require more skilled employees for operation.

8. DATA OUTPUT. The cost of reproducing information must be evaluated in relation to other elements in the system. One system may have a low output cost, but may have high processing and storage costs. The type of output produced is also influential—hard copy output is generally more expensive than soft copy output. Manual output methods are generally more expensive than computerized output.

9. DATA STORAGE. The cost of storing data varies greatly between systems, and it must be evaluated in relation to other relevant factors. These include accessibility, type of storage, and ease of processing. Systems that can store large volumes of data at relatively low cost may involve higher investments in equipment. Others may store less amounts of data, but may have faster accessibility and higher costs.

10. PROGRAMMING AND SOFTWARE. The cost involved in maintaining, updating, and operating a computer program or system software must be assessed during system evaluation. High costs for computer programming and highly trained personnel needed by one system may be offset by the greater number of employees needed to process the same data in another system. Again, these factors must be measured and considered in relation to others.

11. OTHER COSTS. Miscellaneous costs—telephone line charges, duplicating services, consulting fees, lease of backup equipment—must be anticipated and considered during system evaluation.

### Hardware Performance

Performance of system hardware is an important element to be measured and assessed. Speed, reliability, service, maintenance and operating costs, and power requirements are only a few factors that are involved.

1. COMPUTER SYSTEMS. Since computer systems usually represent a major portion of the costs involved in a business system, their efficient operation is of vital importance to the system's overall performance. Several elements in the computer system must be evaluated and compared against the original expectations. These include overall processing time, amount, degree of accuracy, reliability of equipment, maintenance problems, and suitability of output.

Performance of storage devices must be examined in relation to expected results. The system should also be evaluated in terms of the training required by operators and users. Are the training programs adequate and relevant? Do operators have sufficient knowledge to run the machines with little or no difficulty? Do users receive sufficient instructions to enable them to process the data or to perform other operations on the machines?

Total operating costs for the computer system must be evaluated in light of estimated costs and any differences investigated. Table 11.4 is an example of a three-year comparison study of a computer system.

**Table 11.4.** Three-year cost analysis (insurance company claim processing system—based on 3,000 claims per month)

| *Item* | *Old System One Year* | *1st Year* | *New System 2nd Year* | *3rd Year* |
|---|---|---|---|---|
| Equipment Lease | $ 1,450 | $ 3,800 | $ 3,400 | $ 3,200 |
| Maintenance | 1,200 | 800 | 800 | 800 |
| Labor | 44,000 | 22,000 | 22,000 | 22,000 |
| Training | 300 | 4,000 | 3,000 | 2,000 |
| Operating | 6,000 | 8,000 | 2,000 | 1,000 |
| Totals — | $52,950 | $38,600 | $31,200 | $29,000 |

2. OTHER SYSTEM HARDWARE. Other devices and equipment used in the system must also be evaluated and compared against expected performances. These include adding machines, calculators, typewriters, copying machines, and storage devices and facilities.

### Software Performance

Performance of system software must be evaluated. Software includes computer programs, office procedures and routines, manuals, and other documentation related to operating a system. Evaluation includes measuring speed of processing, amount and quality of output produced, accuracy, reliability, and amount of maintenance and updating necessary to keep the software current. Errors often do not show up in software routines until they have been executed several times, and they must be diagnosed and corrected.

The sufficiency of training required by users must be assessed. If present programs are inadequate, their weaknesses must be determined and plans made for their correction. The results of the evaluation should be compared to the expectations specified during system design.

## Productivity

Productivity is a measure of system performance that states the relationship, or ratio, of input cost to output level. The productivity of an entire system, or its parts, can be measured to gain insights into system performance.

The productivity is found by dividing the level of output produced by the system by the input costs. Relatively low input costs and a high output level produce a high productivity ratio. Low productivity is manifest when the output volume is low and input costs are high.

Table 11.5 is a productivity report showing a comparison of costs and output level for an old system being replaced by a new system in an office situation. In this example, the office staff was divided into two sections—one performing the old routines and the other the routines specified by the new system. This arrangement allows for a more accurate comparison of the two systems.

**Table 11.5.** Monthly productivity report

| | *Before Implementation Entire Dept.* | *After Implementation* | | | |
|---|---|---|---|---|---|
| | | *One-Half Dept. Old System* | | *One-Half Dept. New System* | |
| *Category* | *Old System* | *Total* | *Ratio* | *Total* | *Ratio* |
| Number of Employees | 30 | 15 | | 15 | |
| Salaries | $3,100 | $1,550 | | $1,550 | |
| Equipment Costs | $ 600 | $ 300 | | $ 800 | |
| Acceptable Reports Per Person | 420 | 220 | .12 | 860 | .37 |
| Phone Queries Handled | 260 | 140 | .08 | 220 | .09 |

In this example, the productivity ratio for the old system is .12 for acceptable reports produced, and .08 for phone queries handled. The corresponding figures are .37 and .09 for the new system. In this instance, the new system is more efficient and economical.

## Accuracy

Accuracy is a measure of the freedom from errors, or conformity to truth, obtained by a system. Obviously, a high degree of accuracy is a desirable goal, since a low accuracy rate may diminish the utility of an entire system.

Accuracy is usually related to productivity. The error rate usually rises along with the output volume—as speed goes up, the accuracy rate goes down.

Rate of accuracy can be determined in several ways. It may be measured by comparing the results of processing from one system to the results from another. Results produced by a system can also be compared to a known.

The frequency of errors (the number of errors that occur during a given period) gives a measure of system performance. It is manifest in such things as the number of misfiled orders, or misshipped goods. The magnitude of the errors may also be assessed. A large number of small errors may do less harm than one large, gross error.

### Document Integrity

Another measure of performance is the degree of security and control that can be maintained over the documents and records in the system. High document integrity means that records are safe, confidential, and under system control at all times. There is no leakage of vital or private data to outsiders or to those who have no need for the information.

Weighing document integrity is an important factor. High integrity achieved at the expense of raised operating costs may not be practical. But a system with low operating costs and high output that results in many lost records or little control over the dissemination of confidential data may not be acceptable either. A suitable compromise will have to be found.

### Morale

The measure of employees' morale is a criterion of system evaluation. Morale reflects the satisfaction and acceptance that employees feel toward their jobs. The higher the morale, the greater the expected work performance level. Improvements in some areas of a system—output volume, turnaround time, costs—may be achieved, but the changes in procedures and working environment may have a detrimental effect on morale. Such a situation may result in many employee grievances, high absenteeism, discourteous treatment of customers, or other signs of dissatisfaction.

Morale is difficult to measure directly, such as in dollars or minutes on a clock. It is a complex matter involving many psychological considerations, and must often be measured indirectly. Absentee rate or employee turnovers are two indications that can be used.

Table 11.6 is a report comparing absences and late arrivals that occurred before and after installation of a system. It shows that in this situation absenteeism was definitely less after the new system was implemented, indicating an improved level of employee morale.

Manifestations of high morale may sometimes be observed by the analyst —an employee may give extra cooperation during a difficult period or may avert a troublesome situation.

**Table 11.6.** Monthly personnel absenteeism report

| *Reasons* | *Old System* | *New System* |
|---|---|---|
| Absenteeism: | | |
| Valid | 33 | 29 |
| Invalid | 18 | 6 |
| Partial Days | 16 | 4 |
| Late Arrivals | 16 | 6 |

## User and Customer Reaction

The responses of those who use the system or will be affected by it afford another means of measuring system performance. Large numbers of complaints from customers concerning errors or changes in monthly statements would indicate poor performance at one or more points in the process. Fewer complaints might indicate that the new billing procedure was more adequate than the old one. Similarly, requests from other departments for additional information or different kinds of reports could indicate a new awareness of the ability of the computer system to manipulate or restructure data.

Table 11.7 is a report that compares the number and kinds of complaints received from customers under the old system with those received under the new system. The totals have been tallied by quarters of a year and reflect the seasonal variations in the business cycle.

**Table 11.7** Customer service evaluation report—by quarter

| | *Old System* | | | | *New System* | | | |
|---|---|---|---|---|---|---|---|---|
| *Types of Complaint* | *1st Qtr.* | *2nd Qtr.* | *3rd Qtr.* | *4th Qtr.* | *1st Qtr.* | *2nd Qtr.* | *3rd Qtr.* | *4th Qtr.* |
| Lost Shipments | 18 | 12 | 13 | 48 | 47 | 6 | 3 | 3 |
| Late Shipments | 29 | 31 | 16 | 150 | 31 | 12 | 8 | 6 |
| Damaged Goods | 14 | 12 | 12 | 6 | 12 | 8 | 7 | 14 |
| Misfilled Orders | 81 | 43 | 29 | 110 | 65 | 45 | 20 | 19 |
| Misbilled Orders | 16 | 21 | 8 | 13 | 7 | 7 | 23 | 19 |
| Shortages | 3 | 2 | 9 | 1 | 8 | 3 | 4 | 2 |
| Totals — | 161 | 121 | 87 | 328 | 170 | 81 | 65 | 63 |

## EVALUATION OF BENEFITS GAINED FROM SYSTEM IMPLEMENTATION

After performance of the new system has been measured and evaluated, the results are compared against the original expectations determined during systems analysis. These benefits include expected cost savings, increases in output volume and accuracy level, and improvements in turnaround time and customer relations.

If all the preset goals have been reached, system implementation can be considered a complete success. More likely, however, some differences will be found between goals and actual achievements. The degree of disparity should be expressed in quantitative terms: the costs of operating the system are 10 percent greater than anticipated; turnaround time was reduced only 10 percent instead of 25 percent as expected.

The data accumulated during the system evaluation process should be studied and analyzed to pinpoint the reasons why performance did not reach expectations—the format of the source documents led to a high number of errors made when keying the information onto input data records; inadequate preparation of computer operators is slowing down turnaround time of computer processed data and reports; frequent breakdowns of computer machinery is raising maintenance costs 50 percent above the estimated amount.

The next step is to determine whether improvements or adjustments to the system are indicated. The analyst must balance the costs involved in time spent in further system analysis, design, and implementation against the possible improvement of benefits gained.

In areas that are not crucial to overall system performance, or where differences between achievements and expectations are minimal, making improvements may not be worth the additional investment of time and money. If, on the other hand, the deficiencies are in areas of critical importance to the firm or other parts of the system, or the amount of disparity is great, it will most likely be best to return to an earlier point in the system design process to find an improved solution.

Table 11.8 is a report listing the estimated and actual costs for operating a new system. The final totals of the estimated and actual costs are quite close. A disparity in the salaries category is offset by differences in costs for outside contract labor. The analyst would probably want to investigate the reasons for these differences to determine if more adjustments in the system are necessary.

The evaluation process gives the analyst an excellent opportunity to observe the system in operation, and it may lead to additional improvements in system design. Often these improvements were not apparent until after the procedures and activities of the new system were actually performed. Suggestions for improvements and changes may also come from employees, users, customers, managers, and members of the data processing and system analysis departments.

**Table 11.8** Estimated vs. actual cost

| Description | Old System | | New System: Estimated Costs | | New System: Actual Costs | |
|---|---|---|---|---|---|---|
| Salaries: | | | | | | |
| Clerks | 810 hours at $4.50/hour = | $3,645.00 | 300 hours at $5.25/hour = | $1,575.00 | 400 hours at $5.75/hour = | $2,300.00 |
| Managers | 216 hours at $7.00/hour = | 1,512.00 | 25 hours at $9.25/hour = | 231.25 | 45 hours at $8.25/hour = | 371.25 |
| Telephone charges | 16 lines at $22/month = | 352.00 | 12 lines at $22/month = | 264.00 | 12 lines at $18/month = | 216.00 |
| Postage | $75/month = | 75.00 | $100/month = | 100.00 | $125/month = | 125.00 |
| Supplies | $190/month = | 190.00 | $110/month = | 110.00 | $140/month = | 140.00 |
| Outside contract labor | 300 hours at $5.50/hour = | 1,650.00 | 200 hours at $5.50/hour = | 1,100.00 | 50 hours at $5.50/hour = | 275.00 |
| TOTAL — | | $7,424.00 | | $3,380.25 | | $3,427.25 |

## EXERCISES

1. Describe the purpose of system evaluation and why it is important.
2. List some common criteria used to measure system performance.
3. Define "turnaround time."
4. List the major cost elements generally evaluated when reviewing system performance.
5. Define "productivity" and discuss how it is measured.
6. Discuss some of the ways in which employee morale may be observed.
7. Describe the turnaround time analysis report and indicate what kinds of information are generally found in it.
8. Describe a monthly productivity report and indicate what kinds of information are found in it.
9. Describe how the customer service evaluation report is useful in measuring system performance.
10. Define "document integrity" and tell why it is important to the business firm.
11. Develop a system evaluation of one of the systems in your school (registration, grade reporting).
12. Visit a business firm and discuss system evaluation. Describe the documents and procedures used in evaluating a system.
13. Visit a business firm and discuss employee morale with line workers and supervisors.
14. Observe a clerical operation, noting the volume of documents handled and the number of personnel involved. Compare productivity during several different visits.
15. Trace the flow of a document through an organization. Record the number of hands through which it passes and the security measures taken to protect its integrity.

# Chapter 12

# Documenting the System

When systems were less complicated and hardware costs were lower, details on system operations could be easily memorized, or recorded on a few sheets of paper. Questions and adjustments were readily handled by using common sense.

Today, however, the complexity and sophistication of modern business systems dictate a more reliable method of describing a system. Preparing the documentation that comprises this description is the last step in system development. It involves creating, collecting, organizing, storing, citing, and disseminating documents containing information relating to the structure and details of a system. Documentation on a business system or computer program consists of written manuals, system and program flowcharts, text narratives, layouts of input or output records, photographs, and drawings on how a system functions.

These formal descriptions of a system mark its change in status from a project under development to a completed entity, capable of existing without the guidance and control of its creators. It can now be understood, operated, and modified by others as the need arises.

## WHY DOCUMENT A SYSTEM?

The nature, size, and complexity of a system determine the optimum amount and scope of documentation. It is particularly important to document complex computer programs and data processing and communication network systems. The intricate details involved in such systems cannot be left to memory.

The primary advantages of documentation include:

1. COMMUNICATIONS TOOL. Documents, flowcharts, and descriptive material on a system enable analysts and programmers to communicate with each other effectively. System details can be discussed and related to each other on a common ground. Documenting elements of a system graphically reduces the possibility of ambiguous explanations or inaccurate references.

2. FACILITATES TROUBLESHOOTING. Detection and correction of malfunctions or error conditions in a system are greatly aided by adequate documentation. It facilitates analyzing a system and tracing its inner workings. It gives the troubleshooter a basis for comparing the present level of system performance to the normal level indicated in the documentation. Such documentation is as much a diagnostic tool to the systems analyst as is the x-ray machine to the physician.

3. SYSTEM REVISION. Thorough system documentation greatly facilitates revising, changing, or modifying an existing system. It gives an analyst, whether familiar or unfamiliar with a system, a clear picture of how the system operates and how its elements interrelate. Without this information, it would be virtually impossible to change any part of the system without disrupting one or more of the other elements involved.

4. OPERATOR AND PERSONNEL TRAINING. Documentation is a key reference used when designing training programs for the operators and personnel who staff the work stations in a system. It contains the specifications that indicate how a job is to be done, by whom, where, or why. The training director will rely on this information to ascertain how much and what kind of knowledge and skills are needed by employees to perform their jobs adequately.

5. CONSISTENCY. One advantage gained from adequate documentation is the increased level of consistency and uniformity throughout a system. This is especially valuable in a decentralized system where the parts operate in different offices or cities. Good documentation facilitates maintaining consistent application of procedures, policies, and practices in the various offices, units, or departments of a system. It assures that similar demands placed on the system will receive the same responses regardless of what city, branch, or department processes them. It also encourages consistency of action over a long period of time. It helps to avoid the introduc-

tion of minor deviations into established procedures during day-to-day operations.

6. MANAGEMENT TOOL. Documentation provides management with a picture of how a system operates. It facilitates system management and provides much of the background information needed when making intelligent decisions. It is used for planning long-term objectives and goals as well. Without it management would experience great difficulty in coordinating the operation of the elements in a large-scale system, especially if they are scattered geographically. Sound decision-making practices rest upon complete data, facts, and information.

7. ELIMINATION OF REDUNDANCY. The task of documenting a system forces the analyst to take a critical look at the entire system. This scrutiny sometimes uncovers non-essential details and duplication of effort, equipment, and personnel.

8. AUDITING. Most business organizations are subject to auditing and review by many agencies. Complete system documentation enables auditors to identify personnel, records, facts, and data quickly and accurately.

## MAJOR SYSTEM DOCUMENTATION

Particular documents selected for a system depend upon its complexity and purpose. The major pieces of documentation are described below:

### 1. Policy Manual

Policy manuals are documents that state the long-term goals and objectives of the company as perceived by the board of directors, administrators, and top management. These manuals clarify and simplify management's task. They indicate priorities and the hierarchy of corporate values. Policy manuals state policies and guidelines that direct action—not step-by-step detailed procedures on how to carry out a given task. They enable foremen, supervisors, and others to direct subordinates in a manner that is consistent with company policy. Before any specific procedures can be carried out, it must be interpreted in light of company policy. Figure 12.1 is an example of a page from a policy manual. It states the company's sales terms and agreements. Figure 12.2 illustrates a mail use policy.

### 2. Procedure Manuals

Procedure manuals are documents that spell out in detail the steps involved in carrying out a given routine or action. They establish the daily procedures of the company by defining how each task is to be done—always in accordance with stated company policy. The procedure manual is explicit

**Figure 12.1.** Sales terms and policy

All sales accepted by the company will adhere to the following terms and conditions:

1. PRICES. All prices are FOB Los Angeles plant. Prices do not include installation, checkout or maintenance. Title to goods passed to customer when order placed on board common carrier.

2. DISCOUNTS. Invoice due and payable on the 10th of the month following purchase. No discount allowable.

3. CREDIT. Credit will be extended to rated companies. Approval of credit must be made before shipment of goods otherwise goods will be shipped COD.

4. PACKAGING—DOMESTIC SHIPMENTS. Prices include normal packaging and labelling for domestic transit. An extra charge will be made for packaging for export.

5. RETURNS. No goods will be eccepted for return, unless accompanied by our authorization.

6. TAXES. Prices are net. Excise and sales taxes, where applicable, are to be paid by customer.

7. DAMAGED MERCHANDISE. All goods have been inspected when they leave plant. Claims for damage in transit should be made to common carrier.

8. OUT OF STOCK MERCHANDISE. Every effort will be made to fill orders with available merchandise. Items which can be obtained within 30 days of the date of order will be backordered. When received by this company, they will be forwarded to customer under separate packaging.

9. WARRANTIES. No express or implied warranties are made concerning the conditions and use of goods except those stated on the original order.

in detail. It defines each specific step in handling such things as payrolls, financial matters, and returns of merchandise. Figure 12.3 is a page from a procedure manual for an insurance company. It states the steps involved in processing claims.

Procedure manuals are sometimes organized by department. The procedures related to the operation of a given department are gathered together in a single manual and placed in that department for easy reference by personnel.

**Figure 12.2.** Mail and telephone policy

The mail room and switchboard are designed to serve the needs of our customers and employees in carrying out the duties with the company. All employees are requested to adhere to the following policy regarding use of telephone and mail facilities:

*MAIL*

1. Where possible, use first class mail. Identify the originator and room number in the left hand corner.

2. Use only approved company stationery.

3. Type recipient's name, address and zip code.

4. Personal mail cannot be mailed at company expense.

*TELEPHONE*

1. Answer all phone calls promptly. If necessary, transfer the call by informing the caller and signalling the operator. Remain on line until operator has acknowledged.

2. Local calls may be placed from your extensions. Out-of-area calls will be intercepted by automatic dialing equipment. Place out-of-area calls directly through switchboard.

3. When leaving your name and telephone number, also give extension number.

4. Employees are requested not to make personal calls through the firm's switchboard or extensions. Phone booths are available in the rest areas, locker rooms, and personnel department, for use by employees.

### 3. Forms Manuals

Forms manuals are collections of all the business forms used in a system. They specify the layout, size, type of stock, and other relevant details for each form. They list the source of supply, order points, and storage information. They may also include details on how to complete or fill out each form, and information for handling and routing. The manual should also specify any procedures involved if modifications or revisions become necessary.

Assembling all forms into a single manual facilitates the development of a comprehensive forms program. This assures the consistent appearance, quality, and utility of the forms used by a system, reduces costs involved, and builds a better corporate image. Figure 12.4 is an example of a page from a forms manual.

**Figure 12.3.** Procedure manual

*Procedure for processing claims:*

1. Claim form received from agent is reviewed to be sure it is complete and contains all necessary data.

2. If claim form is incomplete, pink "request for additional information" is completed, attached to form and returned to agent. Yellow "out for data" card is completed and filed in cabinet.

3. Completed forms are processed as follows: Green "detail report" is typed using data provided on original claim form. Answer questions accurately and completely. Place the CL number in the box at top of form. Review form to check your accuracy.

4. Submit form to supervisor. When returned, enter necessary adjustments and changes.

5. Photocopy detail form. Send original to payment department, file duplicate in office, send third copy to agent.

6. Complete blue "claim processed card." Indicate disposition of claim. Include CL number. Forward card to supervisor.

### 4. Sales Manuals

Sales manuals define how sales orders are to be written and processed. They outline procedures for such things as closing sales, meeting objections, finding leads, handling financing. Figure 12.5 illustrates how one company guides its sales personnel through a sales manual.

Sales manuals are used by the sales and marketing departments to document their long- and short-term goals. They describe department strategies and indicate which lines of merchandise are profitable and which are not.

### 5. Training Manuals

Training manuals describe the programs that have been set up to prepare personnel to operate the system. They document the level of ability needed to operate each work station, any prerequisite skills, and the details of instructional programs that have been designed to train employees for the various jobs.

Assembling this information into a single manual encourages consistency of training programs, reduces the instances of time spent on teaching skills previously acquired or time spent attempting to teach skills to personnel who have not yet received the prerequisite background. By documenting the situations and procedures that employees have been trained to handle, it allows them to upgrade their level of skills more efficiently to achieve higher job classifications.

**Figure 12.4.** Forms manual

FORM NAME: Request for Confidential Info.

FORM NUMBER: AL612

COLOR: Green

SIZE: 5½" x 8½"

STOCK: 20 l G. # 4 sulphite bond

ORIGINATION: Insurance sales department

DISSEMINATION: Copy to employer

STANDARD ORDER QUANTITY: 5,000

REORDER POINT: 500

APPROVED BY: J. C. Connors 5/12

REQUEST FOR CONFIDENTIAL INFORMATION

The employee below has requested that you, as his employer, provide our firm with the following information. It will be used in processing an insurance policy.

Please complete all items below and return the form to us within 10 days.

NAME ______

ADDRESS ______

______

PHONE ______ BIRTHDATE ______

PRESENTLY EMPLOYED BY ______

ADDRESS ______

______

SUPERVISOR ______ BADGE NO. ______

DATE EMPLOYED ______

SALARY ______ POSITION ______

Has employee been continuously employed during this period? YES ___ NO___. Please explain any interruption in service ______

______

______

**Figure 12.5.** Sales manual

### How to Sell

The following guide lines and sales material have been developed to help you sell in this market. After a discussion of these, this booklet will present "The Case of Miss Dee," demonstrating the application of the information and sales material described. This should serve as a firm foundation for actual sales. A thorough review of the Rate Book For The Pension Series is recommended. This book describes the provisions of our Retirement Annuity contracts and Retirement Income Endowment policies, which *are the only* fixed dollar individual contracts that we now sell under this law. Also, Special Rate Tables for Tax-deferred Annuities, *ORD 31584,* have been printed separately to simplify the calculations necessary in these cases. These tables show results at age 65 per $100 yearly payment based on "P" rates.

1. **For getting an appointment with the school board or employer—sample letters.**

   These specimens illustrated in the Appendix are for making the initial approach to the employer. They may also be used as the basis for a telephone contact. They request an appointment to speak to the employer about the benefits under the law, and should be addressed to the attention of the administrative head of the organization. This approach may not be necessary if a Prudential representative has a personal contact who can arrange such a meeting.

   When you are interested in dealing with school systems, a Directory of Public Schools is very helpful as a prospecting tool. It is usually available through the County Superintendent of Schools, and generally contains the names and addresses of all schools, the names of board officers, and teachers by grade or subject taught.

2. **For talking to the employer.**

   In talking to the employer you may use the information in the initial section of this material. In any event, it is suggested that the following steps be taken:

   a. Acquaint the employer with the benefits to both employee and employer under the law, and how Prudential is prepared to implement the law with our Retirement Annuity contracts or Retirement Income Endowment policies.

   b. Point out that only with the employer's cooperation can the employees take advantage of the law. When dealing with a school system, a board resolution will frequently be required. (We can furnish an illustrative copy as shown in the Appendix of this material, for the board's attorney. It is printed separately as form Ord. 30863.) Either an authorized official must be the applicant for the retirement contracts or a copy of the salary reduction agreement must be furnished with the application or within the same calendar year that the application was written. It is also possible for a school board to remit purchase payments from the proceeds of the salary reduction agreement for an annuity issued on a regular basis provided the nontransferable endorsement is added to the policy. However, such an annuity may not receive qualified dividends.

c. Request permission to talk to the employees in a group to acquaint them with the law.

At the conclusion of the talk there probably will be questions. In the Appendix of this material are some of the more frequently asked questions and their answers. Being familiar with these will give you added knowledge and confidence.

Before leaving the meeting you may wish to distribute a specially designed brochure "Minus = Plus," Ord. 30857, which gives the highlights of the program and a dramatic picture of the possible gains.

Source: "Tax-Deferred Fixed Dollar Individual Annuities," Prudential Insurance Co., ORD 31798 Ed. 10–71 pg. 7.

### 6. Financial Management Manuals

Financial manuals are prepared to guide personnel in the control and movement of funds within the company. They provide a common base for making financial decisions. They define how to handle credits, collections, cost and tax accounting procedures, and how to depreciate assets. They specify such information as when assets should be salvaged or scrapped.

### 7. Organization Manuals

The organizational chart of the firm shows the direct lines of authority and the relationship of such things as offices and subordinates. The organization manual describes these responsibilities and relationships in graphic form, indicating how each role in the firm fits into the overall organization scheme. Figures 2.1 and 2.5 are examples.

## 8. System Manuals

System manuals contain documents that trace the flow of data or goods throughout a system. They provide an effective means of detecting errors or malfunctions in system performance. This manual shows where data originates, where it is processed within the system, and where it is output. It shows the work stations, personnel, and the paths that data and goods follow as they are processed.

System flowcharts and decision tables are the major types of documents used for illustration in this category. Figures 8.4 and 8.5 are examples.

## 9. Consumer Documentation

This category includes any documents prepared for the consumer of a firm's services or goods. They include informative materials, such as shown in Figure 12.6, directions for using services, instructions for filing claims, or any other explanatory text designed to meet customers' needs.

Collecting these documents into a single file allows them to be developed as an efficient and coordinated communications link with the customer. It assures that the documents are adequate, consistent, and comprehensive. It facilitates making revisions or deletions in one or more pieces of documentation without invalidating information or specifications contained in another.

**Figure 12.6.** How to choose and use retail credit

CREDIT AGREEMENTS .....AND CONTRACTS

The Federal Truth in Lending Law requires disclosures similar to the following in retail installment transactions.

**SAMPLE SEARS, ROEBUCK AND CO.**
**REVOLVING CHARGE ACCOUNT AND SECURITY AGREEMENT**

I agree to the following regarding all purchases made on my Sears Revolving Charge Account and Security Agreement.

1. I have the following options each monthly billing period.
   - A. I will pay the entire New Balance within **30** days of the Billing Date shown on the monthly billing statement; or
   - B. I will pay the deferred payment price for each purchase consisting of the cash price and a Finance Charge.
     - (1) The Finance Charge will be computed upon the Average Daily Balance of my account in each monthly billing period.
       - (a) The Average Daily Balance is determined by dividing the sum of the balances outstanding for each day of the monthly billing period by the number of days in the monthly billing period.
       - (b) The Balance Outstanding for each day of the monthly billing period is determined by subtracting payments and credits from the previous day's balance excluding any purchases added to the account during the monthly billing period and excluding any unpaid Finance Charge.
     - (2) **FINANCE CHARGE** will be determined by applying a periodic rate of **1.5%** per month **(ANNUAL PERCENTAGE RATE OF 18%)** to the Average Daily Balance.
     - (3) When the Average Daily Balance for a monthly billing period is $33.00 or less, the **FINANCE CHARGE** for that billing period, at Sears option, will be **50¢** instead of the amount computed above.
     - (4) No Finance Charge will be assessed:
       - (a) In a monthly billing period during which there was no Previous Balance;
       - (b) In a monthly billing period during which payments and/or credits equal or exceed the Previous Balance;
       - (c) On unpaid Finance Charge; or
       - (d) On purchases during the monthly billing period in which they are added to the account.
     - (5) I will pay the deferred payment price for all purchases in monthly payments within **30** days from each monthly Billing Date according to the following schedule:

| If the New Balance Is: | The Minimum Payment will be: | If the New Balance is: | The Minimum Payment will be: |
|---|---|---|---|
| $ .01 to $ 10.00 | Balance | $350.01 to $400.00 | $30.00 |
| 10.01 to 200.00 | $10.00 | 400.01 to 450.00 | 35.00 |
| 200.01 to 250.00 | 15.00 | 450.01 to 500.00 | 40.00 |
| 250.01 to 300.00 | 20.00 | Over $500.00 | 1/10 of New Balance |
| 300.01 to 350.00 | 25.00 | | |

       I have the option of paying more than the minimum payment each month.
     - (6) If I fail to pay any Minimum Payment when due, Sears may declare my entire balance due and payable.
     - (7) Sears shall retain title to merchandise purchased under this agreement until paid in full. Each payment shall be applied to merchandise and services as follows: first to unpaid Finance Charges; then, as to items purchased on different dates, the first purchased shall be deemed first paid, as to items purchased on the same date, the lowest priced shall be deemed first paid.
     - (8) I have the right to pay my entire balance in full at any time without incurring a subsequent Finance Charge.
2. I agree that the Finance Charge and other credit terms will be determined by the law of my State of residence. If I change my State of residence, I will inform Sears, and Sears will provide me with a new agreement containing the Finance Charge and other credit terms applicable to my new State of residence.
3. Sears is authorized to investigate my credit record and report to proper persons and Bureaus my performance of this agreement.
4. Sears waives the right to retain or to acquire any lien arising solely by operation of law in real property used or expected to be used as my principal residence. This provision is not applicable to judgment liens.

**NOTICE TO BUYER: (1) DO NOT SIGN THIS CONTRACT BEFORE YOU READ IT OR IF IT CONTAINS BLANKS. (2) YOU ARE ENTITLED TO A COPY OF THIS CONTRACT. KEEP IT TO PROTECT YOUR LEGAL RIGHTS. (3) YOU HAVE THE RIGHT TO PAY IN ADVANCE THE FULL AMOUNT DUE.**

**RECEIPT OF A COPY OF THIS SECURITY AGREEMENT IS ACKNOWLEDGED.**

Accepted:
SEARS, ROEBUCK AND CO. By ____________

____________________ *(Customer's Signature)*    Date ________ Store No. ________

8

ACCOUNT NUMBER | | | |

**SAMPLE SEARS, ROEBUCK AND CO.**
**DISCLOSURE STATEMENT**

Sales Check No. ____________ Date ____________ 19____

☐ Easy Payment Plan

☐ Modernizing Credit Plan

| DESCRIPTION OF MERCHANDISE |
|---|
| |
| |
| |
| |
| |

| OFFICE USE ONLY (Code 4 Sales) | |
|---|---|
| No. of Months | Monthly Payment |
| | |

| | | | | |
|---|---|---|---|---|
| CASH PRICE | | | | |
| CASH DOWN PAYMENT | | | | |
| UNPAID BALANCE OF CASH PRICE — AMOUNT FINANCED | | | | |
| **FINANCE CHARGE** | | | | |
| DEFERRED PAYMENT PRICE | | | | |
| TOTAL OF PAYMENTS—THIS SALE | | | | |

This purchase is payable in installments pursuant to my Sears Easy Payment Plan—Modernizing Credit Plan Retail Installment Contract and Security Agreement.

Beginning ____________, I will pay $________ per month for ________ months and a final monthly payment of $________ until the amount financed and the finance charge for this purchase are fully paid.

If the **FINANCE CHARGE** exceeds **$5.00**, the **ANNUAL PERCENTAGE RATE** is [   %]

In accordance with my Sears Easy Payment Plan—Modernizing Credit Plan Retail Installment Contract and Security Agreement, a subsequent purchase may change the number and amount of my monthly payments, the amount of the Finance Charge and the Annual Percentage Rate of this purchase. Any such change will appear on my next monthly billing statement.

A copy of my sales check is attached hereto and incorporated by reference. Ownership of the merchandise described in such attached sales check remains in Sears until paid for in full.

If I pay in full in advance, any unearned finance charge will be rebated under the Rule of 78.

**Note:** *Samples of credit agreements, contracts and statements of accounts appearing in this booklet are representative of those in general use in most areas. There may, however, be some variance in your locality due to state and local laws and the policy of individual merchants.*

9

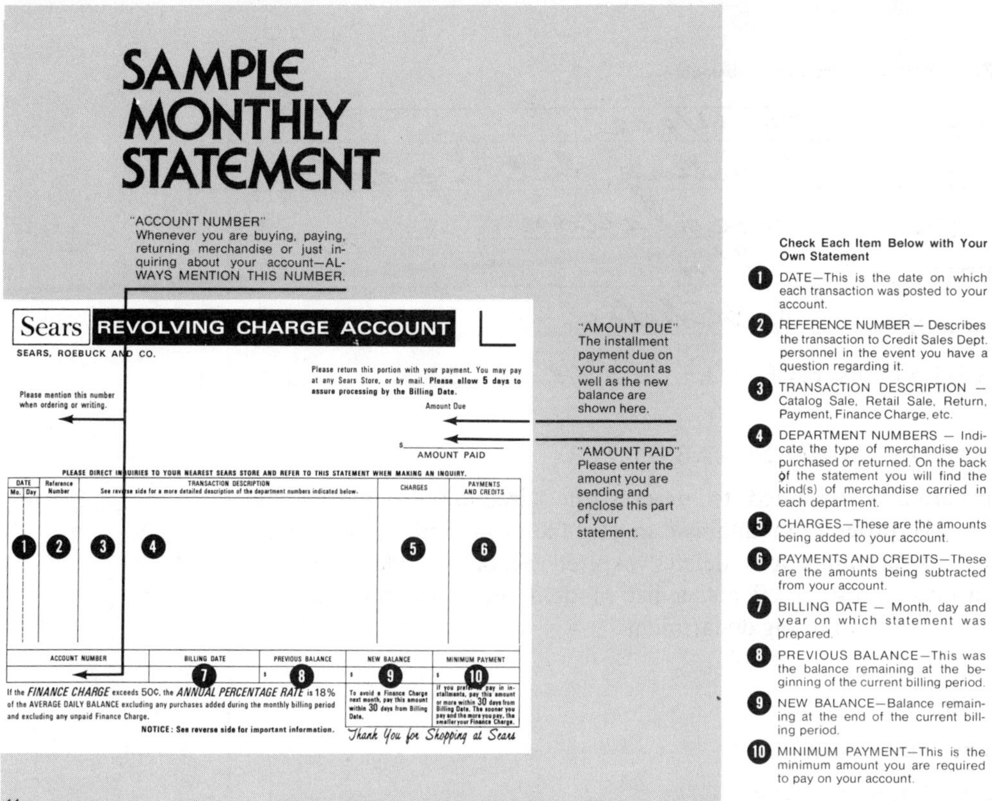

# SAMPLE MONTHLY STATEMENT

"ACCOUNT NUMBER"
Whenever you are buying, paying, returning merchandise or just inquiring about your account—ALWAYS MENTION THIS NUMBER.

Sears REVOLVING CHARGE ACCOUNT

SEARS, ROEBUCK AND CO.

Please mention this number when ordering or writing.

Please return this portion with your payment. You may pay at any Sears Store, or by mail. **Please allow 5 days to assure processing by the Billing Date.**

Amount Due

$ ________ AMOUNT PAID

"AMOUNT DUE"
The installment payment due on your account as well as the new balance are shown here.

"AMOUNT PAID"
Please enter the amount you are sending and enclose this part of your statement.

PLEASE DIRECT INQUIRIES TO YOUR NEAREST SEARS STORE AND REFER TO THIS STATEMENT WHEN MAKING AN INQUIRY.

| DATE Mo. Day | Reference Number | TRANSACTION DESCRIPTION See reverse side for a more detailed description of the department numbers indicated below. | | CHARGES | PAYMENTS AND CREDITS |
|---|---|---|---|---|---|
| 1 | 2 | 3 | 4 | 5 | 6 |

| ACCOUNT NUMBER | BILLING DATE | PREVIOUS BALANCE | NEW BALANCE | MINIMUM PAYMENT |
|---|---|---|---|---|
| | 7 | $ 8 | $ 9 | $ 10 |

If the *FINANCE CHARGE* exceeds 50¢, the *ANNUAL PERCENTAGE RATE* is 18% of the AVERAGE DAILY BALANCE excluding any purchases added during the monthly billing period and excluding any unpaid Finance Charge.

**NOTICE: See reverse side for important information.**

To avoid a Finance Charge next month, pay this amount within 30 days from Billing Date.

If you prefer to pay in installments, pay this amount or more within 30 days from Billing Date. The sooner you pay and the more you pay, the smaller your Finance Charge.

*Thank You for Shopping at Sears*

14

**Check Each Item Below with Your Own Statement**

1. DATE—This is the date on which each transaction was posted to your account.
2. REFERENCE NUMBER — Describes the transaction to Credit Sales Dept. personnel in the event you have a question regarding it.
3. TRANSACTION DESCRIPTION — Catalog Sale, Retail Sale, Return, Payment, Finance Charge, etc.
4. DEPARTMENT NUMBERS — Indicate the type of merchandise you purchased or returned. On the back of the statement you will find the kind(s) of merchandise carried in each department.
5. CHARGES—These are the amounts being added to your account.
6. PAYMENTS AND CREDITS—These are the amounts being subtracted from your account.
7. BILLING DATE — Month, day and year on which statement was prepared.
8. PREVIOUS BALANCE—This was the balance remaining at the beginning of the current billing period.
9. NEW BALANCE—Balance remaining at the end of the current billing period.
10. MINIMUM PAYMENT—This is the minimum amount you are required to pay on your account.

15

Courtesy of Sears, Roebuck and Company.

## 10. Computer Program Documentation

The expanding use of computers within data processing increases the need and importance of including complete descriptions of computer programs in system documentation. This information serves as a major means of communication between programmer, manager, and computer operator.

The important elements in the program documentation file are described below. Computer programs that are relatively simple will require only a few of these items for adequate description. Those that are more complex may require an extensive documentation file containing most of these items. All the information regarding a particular program is collected into a single file. Copies of this file, or portions of it, are made available to management, the systems department, operators, programmers, and other users.

a. Abstract. The abstract is a short, written description that summarizes the function of the program, the logic it follows, and the general routines available. Figure 12.7 is an abstract for a program that updates a mailing list.

b. Flowcharts. Program flowcharts are included to illustrate program flow. They are graphic, visual descriptions of the steps followed during program execution.

**Figure 12.7.** Computer program abstract

PROGRAM NO.: TR 61/R02

PROGRAM NAME: Mailing list update

DEPARTMENT: Circulation

DATE WRITTEN: 7/14

APPROVED BY: J. White

DOCUMENTATION FILE NO.: 1066

PROGRAM DESCRIPTION—ABSTRACT:

This program is designed to process the circulation department's mailing list. The program maintains a master file of all paid up accounts. It merges new accounts and deletes expired accounts. The program prepares a current list of clients, a list of delinquent accounts, and a renewal list for the marketing department.

ENVIRONMENT:

1. Language: BASIC
2. System: Interdata 32K
3. I/O Devices: Card reader<br>Line printer<br>CRT

c. TEXTUAL NARRATIVE. A written narrative describing the logic followed in coding the program is also included. It explains, in detail, the algorithm used, branches, options, and formulas used in manipulating the data in the program.

d. PROGRAM LISTING. A listing of the program statements, showing the coded instructions in the sequence suitable for execution, is an important item in the file.

e. RECORD LAYOUT. A sketch or outline illustrating the layout of the input and output records is prepared. A detailed description of the data fields, codes, and abbreviations may be included. Figure 12.8 is an example.

f. OPERATING INSTRUCTIONS. Instructions to the operator on how to run the program, and what branches and routines are available are included in the documentation file. These instructions also indicate any error messages and program reentry routines, if available.

Figure 12.8. Record layout

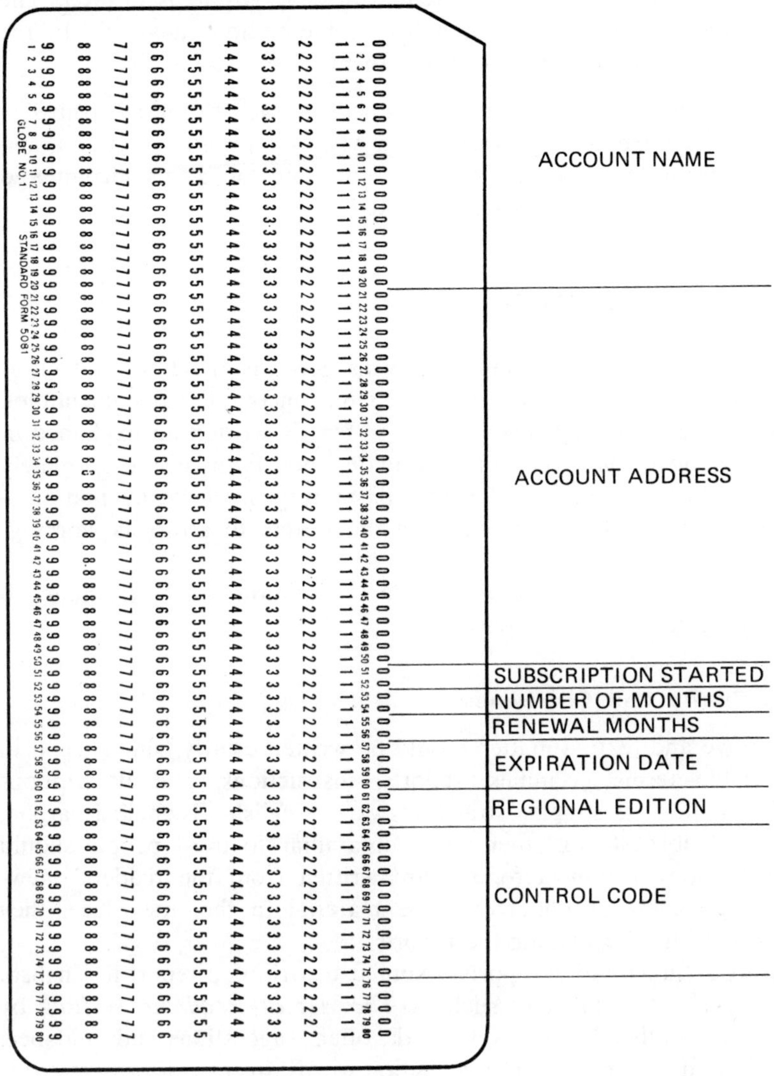

ACCOUNT NAME

ACCOUNT ADDRESS

SUBSCRIPTION STARTED

NUMBER OF MONTHS

RENEWAL MONTHS

EXPIRATION DATE

REGIONAL EDITION

CONTROL CODE

GLOBE NO. 1 STANDARD FORM 5081

g. Options. Many programs contain branches or options which the operator can select to produce different forms of output as desired. The documentation file should contain a list of the options available in the program, and instructions on how to use them.

h. Run manual. A history maintained on each program is filed in the program documentation. This indicates when the program was run, and any changes or modifications made in it. It also notes problems encountered or error conditions that have been corrected.

## GUIDELINES FOR PREPARING DOCUMENTATION

The usefulness of documentation is only as good as the time and effort expended in its preparation. Hurriedly written or incomplete documentation has limited value. Some documentation is so tersely written, or explained in jargon so technical, that it can only be understood by someone intimately familiar with the system. This defeats the basic purpose of documentation—to allow the individual *not* familiar with the system to grasp its workings quickly and easily.

The preparation of adequate system documentation will be facilitated if the following guidelines are followed:

### Maintain Clarity and Conciseness

All descriptive and textual matter should be written clearly, simply, and in understandable terms. Manuals, reports, instructions, and descriptions should be written in simple, straightforward English. Technical jargon, engineering doubletalk, or other terms unfamiliar to most people should be avoided. Good technical reports are written from the readers' viewpoint—if they knew or understood the material in the first place, they probably would not be reading the reports.

Preparing documentation reports is similar to writing a report in English. A good English style manual, such as *The Art of Readable Writing* by Rudolf Flesch, Collier Books, New York, offers suggestions and guidance. Short words and sentences and direct action words, instead of passive terms and long, complex sentence structure, will increase clarity and interest. Short paragraphs and the liberal use of headings and subheads will help to outline textual material and will facilitate referencing data and subjects. Outline the material before writing it. Organize topics carefully and place them in logical order.

### Use Graphics Liberally

"A picture is worth a thousands words" is a well-known proverb particularly applicable to preparing documentation. Use plenty of diagrams, drawings, and artwork to illustrate concepts that are more easily explained with graphics than with written text.

Use these visual techniques to reinforce textual discussions. A series of steps described in text form and illustrated with block diagrams or flowcharts will more quickly be understood and learned than those without graphics.

A set of glossy photographs showing how to do a given operation is often much more effective than a long written treatise. For example, photographs can be used to illustrate various types of damaged goods much more effectively than can a written description. Fully completed sample forms will illustrate the correct method of entering each piece of data much better than will descriptive text alone.

### Relate Pieces of Documentation

The items in the documentation file should be designed to relate to each other in an organized, structured way. Outline or summarize the different documents included in the file. Number paragraphs for easy reference. Cross reference the various documents for clarification. Organize items so that each relates only to a specific area or topic. Don't describe forms routing in the middle of a description on training typists.

### Emphasize Data Flow

Organize and design all documents to facilitate tracing the movement of data throughout the system. Emphasize the relationship of each particular piece of documentation to the points of processing and output.

### Keep the Reader in Mind

Do not make assumptions regarding the knowledge or information possessed by the reader. Take the time to explain preliminary matters or to present the prerequisite knowledge.

Define all terms carefully. Use examples and illustrations liberally. Build from the simple to the complex. Do not begin a detailed technical description unless you are sure your reader is ready to follow it.

Use introductory paragraphs and summaries to explain what will be covered in the section. Follow up the presentation of the material with a summary to reinforce important points.

## DISTRIBUTION OF DOCUMENTATION

A definitive statement on how documentation should be distributed, to whom, and where it should be stored is difficult to make. It should, of course, go to those who will need it. It should be available at the times, places, and at the level of detail appropriate to that particular usage.

Some items of documentation are historical in nature and will be kept in a permanent file not readily accessible to system users. Other items will be constantly available—near machines, work stations, telephones.

Some documentation is prepared for the exclusive use of the systems analyst and is referred to only when making major changes in the system. Other items are prepared for use by management. Policy and procedure manuals may be given to department heads, foremen, or managers.

Some systems analysts organize three-ring binders containing the related documentation for each department in the firm. Each binder, clearly labelled, is placed in the appropriate department to document the established routines and procedures that are performed there. The contents should always be kept current, and any changes made in it should be brought to the attention of the proper supervisor.

## REVISION OF DOCUMENTATION

Documentation is of value only if it reflects the present and complete status of a system. An outdated piece of documentation will probably cause more confusion than anything else. An organized, systematic review of all documentation should be undertaken periodically to keep its contents current. Records should be maintained on any changes made, indicating which documents were involved and the text of the revision. Notification of the changes should be promptly brought to the attention of the appropriate personnel. Outdated material should be removed from the documentation or placed in a separate file.

## LIMITATIONS OF DOCUMENTATION

A strong case can be built describing the advantages of comprehensive documentation on organizational policies and procedures. However, documentation can also become a means of evading responsibility within a firm. For some employees it becomes the end in itself, rather than the means of achieving the firm's objectives. They view documentation as sacred writings that cannot be changed or improved, and as a protective screen behind which to hide from their responsibilities for performing a job effectively.

There is some support for the position that only broad policies should be stated and that responsibility for carrying out these policies should be delegated to appropriate individuals in the organization. This view contends that specific details and guidelines on system operation should not be included in the documentation, but left to the discretion of the individuals in charge. It is their task to structure the procedures that implement the goals of the organization.

A further argument against putting policies in writing is that they become rigid. Some students of organizational theory believe that policies should not be codified for the simple reason that this limits the flexibility of a firm and discourages innovation and adaptation to a changing environment.

There is some truth in both positions. Effective implementation of system documentation falls somewhere between the two extremes of an ad hoc verbal policy and an extensive detailed codified statement.

## EXERCISES

1. Summarize the need for documenting a system.
2. List six reasons why system documentation is important.
3. List six major pieces of system documentation.
4. Describe the information contained in an organization's policy manual.
5. Describe the information contained in an organization's forms manual.
6. Describe the information contained in an organization manual.
7. List the major pieces of information contained in a computer program documentation file.
8. List the major guidelines to follow when preparing system documentation.
9. Describe the writing style and rules that should be followed in preparing manuals, reports, system instructions, etc.
10. Visit a business firm and collect several pieces of system documentation. Describe the function of each.
11. Evaluate the quality of the system documentation obtained above, using the criteria discussed in this chapter.
12. Rewrite or modify the documentation obtained above to improve its content, delivery, comprehensiveness, etc.
13. Write a short policy manual on some phase of a school activity.
14. Prepare an organization chart of your campus administration.
15. Visit a computer center and observe the kinds and types of documentation prepared on their system and programs.

# Chapter 13

# Systems Analysis Tools

Quantitative, scientific tools have been developed by systems analysts to help them to understand, predict, and manage the operation of business systems. The application of these techniques is sometimes referred to as operations research, or just OR. Operations research involves applying the laws of logic, statistics, mathematics, and science to the solution of business problems, such as the allocation of physical resources, project management, inventory control, and business forecasting. Operations research departments have been organized in many companies to improve management's ability to make effective decisions.

## HISTORY OF OPERATIONS RESEARCH

Operations research dates back many years. Frederick W. Taylor, often called the father of scientific management, developed some early OR tools in the late 1880's. He was a pioneer in determining the optimum allocation of resources for a given task.

Frank B. and Lillian M. Gilbreth developed the "efficiency expert" approach early in the twentieth century. In the 1930's Elton Mayo experimented with diversified work environments and employee motivation studies. The goal of these early analysts was to find a way to produce a good or service at the lowest cost.

World War II brought a new emphasis, urgency, and goal to operations research. The United States was faced with the massive job of building a national defense system with a limited amount of resources (personnel, money, time). The problem was how to determine the optimum allocation of resources.

To aid in this task, efficiency experts were called in to help defense plants and the military improve performance. They relied on the OR techniques then in existence to find the most efficient mixture of bombers, tonnage, strikes, and personnel. Other investigators began working on the problem of manufacturing the greatest amount of war material in the least period of time, and with the minimum expenditure of people and money.

The construction of liberty ships during World War II by Henry Kaiser is a notable example. With careful planning, ordering of materials, and scheduling of work loads and shifts, he was able to reduce the number of days it took to build a liberty ship from six months to only a few days.

The development of the modern electronic computer raised the science of operations research to a new dimension and level of sophistication. Techniques with names like PERT, CPM, sampling, forecasting, mathematical modeling, linear programming (LP), and queuing theory were developed and have become an indispensable part of the systems analyst's resources.

## ADVANTAGES AND LIMITATIONS OF OPERATIONS RESEARCH

Applying the laws of mathematics, logic, and the scientific method to systems problems yields certain advantages. These sophisticated techniques increase the reliability and accuracy of the analyst's strategies and recommendations. They extend the range and complexity of problems that may be investigated. They improve the analyst's ability to make predictions and forecasts and manage complex networks of people, machines, facilities, and resources.

One of the most important advantages gained from the use of these systems analysis tools is that the data, information, and recommendations prepared for management are more valid, precise, and comprehensive. This greatly increases their value in the decision-making process.

But, of course, operations research techniques have their limitations too. They cannot solve all problems or eliminate all chances of error. Any results obtained from their use are only as accurate as the information fed into them and the skill used in their execution. These techniques are somewhat sophisticated—sometimes too sophisticated for the problem being tackled. Using operations research techniques to solve some problems may cost more time, money, and effort than the problem situation did in the first place.

In addition, the analyst must always be aware of the tendency to overrate the potency and superiority of these techniques. In many cases, using common sense or knowledge gained from experience will yield similar results—and at a far lower price.

## TYPES OF SYSTEMS

The type of technique used to analyze a problem situation depends somewhat on the kind of system under study. Depending upon whether or not their outcomes and behavior can be predicted with certainty, business systems fit into two categories:

### Stochastic Systems

Systems in which the output, or results, cannot be determined precisely, but only guessed at, are called probabilistic, or stochastic, systems. They involve elements of chance or probability.

A flight reservation system is an example of a stochastic type. The airline operator cannot know for certain, in advance, how many individuals will purchase tickets for a given flight. An estimate of the total can be made, based upon probability or past experience, but the actual number of seats sold for the flight remains unknown until the event occurs.

### Deterministic Systems

Other systems are deterministic—the result or output can be predicted with certainty, in advance. For example, a clock manufacturer knows with certainty how many gears, jewels, dials, bridges, or hands will be required to assemble a fixed number of watches. The number of required components will not change from time to time. It will always remain proportional to the number of watches being assembled.

Output of deterministic systems can be predicted with relative ease and accuracy by measuring the input or demands placed upon the system. For this reason, deterministic systems are easier to manage than are stochastic systems.

Most business systems have characteristics of both types. Some elements are known for certain; others behave stochastically and must be estimated. Figure 13.1 illustrates the two categories. If a major portion of a system is stochastic, the entire system must be treated stochastically.

For example, suppose an airfreight carrier agrees to ship a certain number of packages for a customer each day. The number, weight, and destination of the shipments never change. But, in addition, the airfreight carrier also provides service to those whose daily shipments vary. The result is a system with both deterministic and stochastic elements. In allocating resources (number of planes, pilots, flight schedules, package handlers), the shipper

Figure 13.1. Stochastic vs. deterministic systems

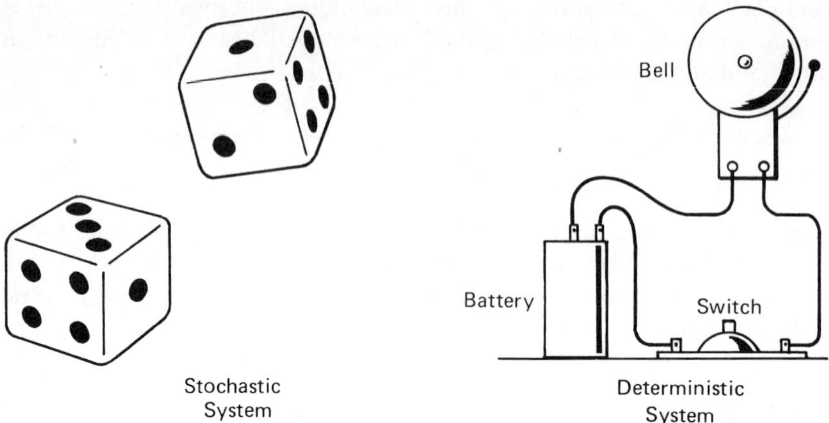


must assume that the entire system behaves stochastically and must estimate the amount of resources needed for each day.

## MAJOR OPERATIONS RESEARCH TOOLS AND TECHNIQUES

The techniques developed by operations research fall into several categories: quality control, allocation of resources, forecasting, and time schedules and planning. Many involve complex mathematical equations and formulas. The description of these techniques, as presented in this text, ignores the details of solving the mathematical formulas. Instead the techniques are described in terms of when and where they are used and what kinds of problems they can solve. Statistics and advanced systems analysis texts should be consulted for a more complete description of these techniques.

The major tools in use today are:

### Sampling Methods

Sampling techniques are widely used in business systems to learn the characteristics of a group of people or items without examining each individual member in the group. These techniques are used in such areas as conducting surveys to study public opinion and employee attitudes, maintaining quality control of manufactured items, measuring time details, and checking inventory levels.

In sampling, only a part of the group is inspected or studied, and the information gained is assumed to be true of the whole group. This method is used mainly in three business situations: 1. where tests on manufactured items for durability or defects result in the destruction of the item; 2. in decision-making activities where conclusions must be made before all the answers and data have been collected and made available; 3. where the group under study is very large and it would be too expensive or time consuming to examine each individual specimen.

Statisticians have long made use of the principles of sampling, and have developed several reliable ways of selecting samples that are truly representative of the larger group. They use the term *population* to refer to the entire group and *sample* to refer to a selected number of units or elements drawn from the group for study. (See Figure 13.2.) The size of the sample need not be large to produce accurate results; rather it must accurately reflect the characteristics of the population being studied.

Following are the different types of sampling:

1. SIMPLE RANDOM SAMPLING. In simple random sampling, all members of the group have an equal chance of being selected for the sample. This method can be used in any situation where all the elements in a group (or numbers representing the elements) are available for selection at the same time. It would be used to sample the kinds of complaints being received by an organization, without reading each individual complaint.

**Figure 13.2.** Selecting a sample from a population

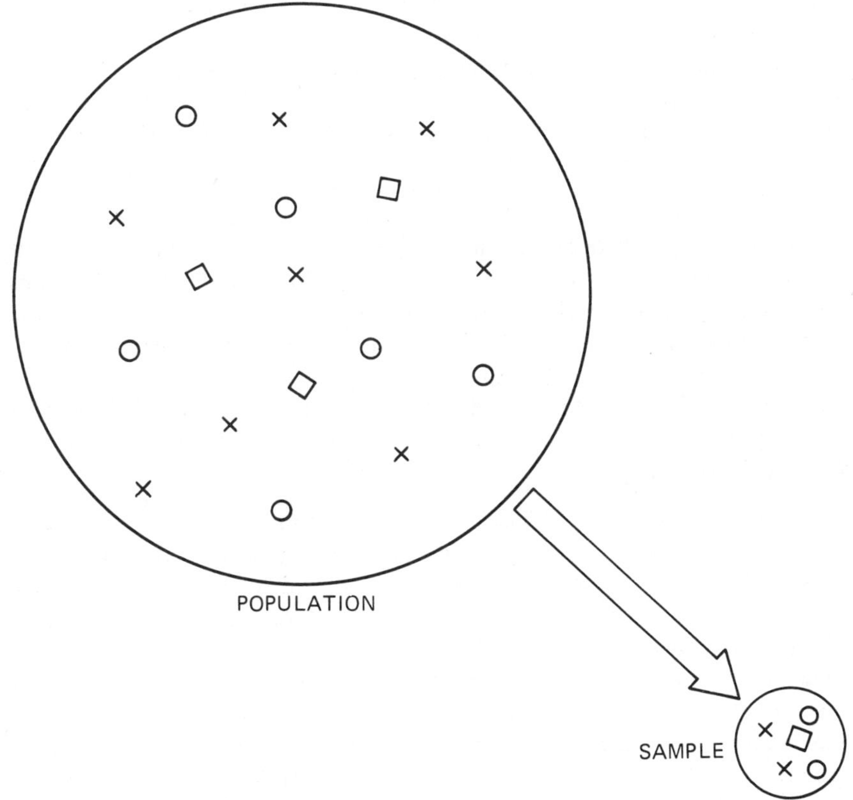

It would be used in quality control practices—one or more specimens would be selected at random from a larger group of items for testing. In situations involving very large populations, this arrangement is impractical.

2. SYSTEMATIC SAMPLING. In this arrangement, samples are selected from a group according to an organized pattern. For example, every fifth member of the group will be selected, or every tenth. Usually the first element in the sample is selected in a random manner. The rest of the elements will be selected according to the systematic pattern. For example, if the pattern is every fifth member and the first element selected (at random) was number 9, then the second element would be number 14, the third, number 19.

3. STRATIFIED SAMPLING. This type of sampling is used to obtain a more accurate representation of each subgroup in a population than usually occurs with a simple random sampling. For example, suppose a survey was planned of all employees in a firm, and it was important that each department be represented in the sample. The elements in the sample could be selected at random, but the analyst could not be absolutely sure that each department was represented. If, on the other hand, all employees in the firm (the population) were first divided into departments (stratified) and a few selected at random from each strata, the analyst could be sure that each group was represented in the final sample.

In this instance, a smaller number of samples would produce more precise and accurate representation than would simple random sampling. This method has even more value when the cost for selecting and examining each sample is an important factor.

4. CLUSTER SAMPLING. This method, also called area sampling, is used when a large geographic area is involved. The population is first divided into small areas or groups. Then a certain number of these groups are selected by systematic or simple random sampling methods. The final sample is chosen from this selection of groups by systematic or simple random sampling.

Cluster sampling is often used for marketing and political surveys. It is less accurate than the others described, but may be more convenient and economical to conduct.

5. QUOTA SAMPLING. This method of sampling is structured so that a specified number of representatives from each group will be included in the final sample. For example, the analyst may be told to interview five members from the sales department, two from the clerical staff, and two from the shipping department.

Business systems analysts use the various sampling techniques to monitor the accuracy of office procedures, data processing activities, computer generated output, and other areas. The systematic sampling method might be applied to measure time delays encountered on the telephone switchboard at various times throughout the day. The analyst might place a call every 15 minutes and record the time that elapses before the call is processed. Quality and accuracy standards of reports produced by the stenographic pool might be monitored by a careful review of a sample selected at random from the day's output.

### Allocation of Resources

One of the common problem situations in system design is determining how to allocate a fixed amount of resources to obtain the optimum return. This situation arises whenever there are alternative ways of combining limited resources. The systems analyst must determine the most effective combination to optimize profits, reduce time delays, or serve the maximum number of customers.

As an example, consider a telephone system that must cope with loads that vary throughout each day and from one day to another. The systems analyst must determine the most efficient and economical arrangement of operators, machines, lines, and equipment to handle the changing loads.

A system designed to serve the peak loads at all times would assure that no customer experienced any delays in getting the dial tone or placing operator or information calls, regardless of when the call was placed. But since the system does not always operate at peak capacity, a large portion of the system elements and personnel will be idle at one time or another. Unused resources represent an economic loss to the company.

This arrangement would, therefore, require a large expenditure in money, lines, equipment, and personnel. To meet the minimum needs of the customers would also be unacceptable. The ideal solution would lie somewhere in the range of employing less than a full complement of operators, staggered so as best to meet the varying needs, and to provide less than a maximum number of lines and equipment. This arrangement would save cost. It would also mean that at certain peak periods some customers would experience delays or reductions in quality of service. Situations such as this always involve compromise—the problem facing the systems analyst is to find the *best* compromise.

1. LINEAR PROGRAMMING. Linear programming, also known as LP, is a mathematical technique developed to determine the best mixture of components. It is invaluable when dealing with complex situations involving many variable resources.

To use it, all the elements that are involved, as well as the expected results, are expressed quantitatively. Then these values are tested in various combinations until the one that most closely meets the stated objective is discovered.

Examples of expected results include "must handle a minimum of 100 queries per hour"; "each unit must cost less than 67¢"; or "find the arrangement that produces the least amount of errors." Resources to be allotted would be expressed as cost per personnel hour worked, cost of materials, number of machines available, total amount of money that can be allotted to the situation, number of people that must be served, and so on.

It is assumed that one or more of the resources are limited, and that a change in the cost or quantity available of one will result in a proportionate change in the others.

While linear programming can be performed mechanically, programming the computer to perform the calculations has greatly expanded the number and range of variables that can be incorporated and the complexity of the problems that can be solved.

2. QUEUING, WAITING LINE THEORY. Another aspect of systems design, related to allocation of resources, is estimating in advance how much load will be placed on a given system at any one time. The analyst must be able to predict what the requests for service will be in a system where the requests fluctuate throughout the day.

If insufficient personnel are available to handle requests, waiting lines (queues) develop. If the delay is long enough, customers or employees get discouraged and find other stores, or means of solving the problem. If too many personnel are available to handle the load, too much time is spent on unproductive activities, and operating costs go up.

Queuing theory techniques are used to schedule assignment of resources so that they can efficiently process the requests for service that arise during any given time period. This scheduling must be flexible enough to handle not only the average number of expected requests, but also an unexpected increase or decrease in load, without involving either an economic loss or excessive waiting time.

The analyst uses two mathematical techniques to estimate the expected and unexpected loads. A study of the average number of requests handled, based on the law of averages, indicates the expected normal load. The theory of probability is used to estimate the unexpected load. Briefly, this theory indicates what the chances are for an event to occur. The analyst uses it to calculate such information as how often twice as many customers as expected might appear for service at the same time.

Based on the pattern that evolves from this study, the analyst schedules the resources so as best to meet the requests that will probably be placed on the system.

Examples of the types of situations that involve fluctuating waiting lines are bank teller windows, checkout stands at markets and department stores, stenographic pools, file inquiry systems, and sales floors.

Figure 13.3 is a chart that graphs the size of the queues which develop on a weekly basis in the classified advertising department of a newspaper. It shows that the queues are largest just before the deadlines for entering ads for each major edition, and shortest just after the deadlines. The classified advertising department must have a system designed to handle a load that varies, not only from day to day, but from week to week as well.

**Figure 13.3.** Classified ad queuing

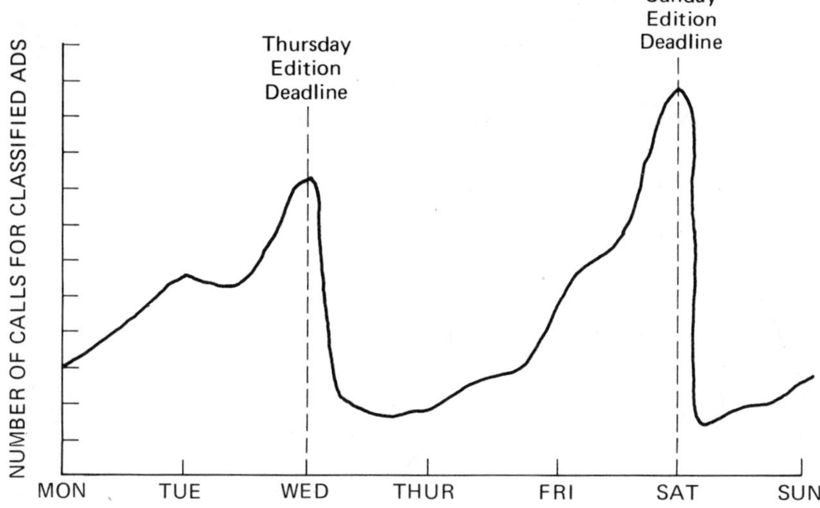


### Quality Control Techniques

In business systems, the analyst often has the responsibility of monitoring the error rate for a given process, indicating when the rate has gone beyond acceptable standards, and instituting measures to correct the problem. This involves monitoring the various office and data processing procedures to assure that the quality control standards set for those processes are being maintained. These procedures include such activities and output as letters, reports, documents, invoices, calculations, entries in records, and computer-generated output.

To examine each specimen produced would be slow, costly, and inefficient. So the analyst relies on random sampling to select a sample for inspection. The size of the sample depends on how closely the error rate of the sample conforms to the standard for that process.

One common quality control method is to continue sampling at a low level as long as the error rate is below or at the standard. If the error rate goes up, the size of the sample rises also—a larger percentage of the specimens being produced are carefully examined. If the cause of the error cannot be found, and the error rate continues to climb, an even larger proportion of the specimens is examined. This process continues until the cause has been found and corrected.

In occasional instances where difficulties still occur, the sample may become as large as one in five, or perhaps include all the units produced. When the condition has been corrected and the error rate begins to drop, the size of the sample falls proportionally. When the error rate has again reached the standard for that process, the sample size is again reduced to the original maintenance level. Figure 13.4 illustrates the relationship of

**Figure 13.4.** Error rate vs. size of sample

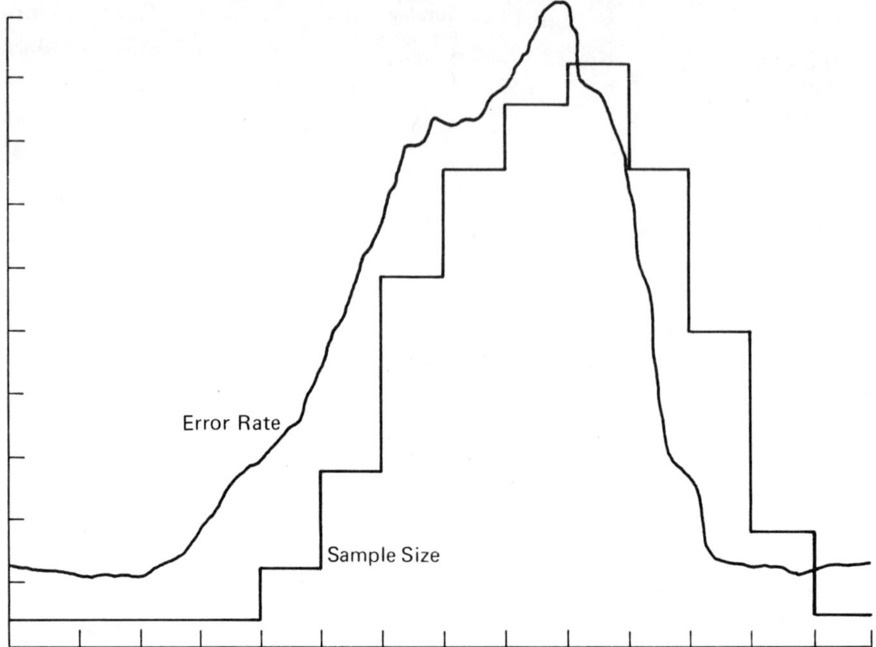


the error rate for a process and the size of the sample taken for examination.

## Forecasting

Forecasting is a technique used to make predictions of future events based upon past performance. The systems analyst is often asked for advice or information regarding future considerations, demands, costs, or output of a system, based upon present available knowledge and data.

LINEAR REGRESSION. Linear regression analysis is an important statistical tool in forecasting. Based upon past performance, this mathematical formula will extend a given line, or sequence of events, into the future.

To use it, all conditions relevant to the area under study are expressed quantitatively. These include costs of personnel, equipment, material, time considerations, output rate, and error rate. These values are manipulated by the formula to produce an estimate of what results will be at a specific time in the future.

Computer programs, written to solve this formula, have increased the complexity of the situations that may be forecast and the accuracy of the results. In real life business situations, however, many factors that cannot be reduced to numbers also influence a given trend. Changes in competition, demand, supply, new inventions, and political and legal factors must also be considered.

The task of the systems analyst is, therefore, to study all of the unpredictable, as well as the predictable, variables affecting a given condition and, by utilizing the appropriate mathematical techniques, to make predictions that are as accurate and comprehensive as possible.

Examples of situations where forecasting would be used include predicting future sales of a firm, a rise in employee salaries, an increase in the number of reports needed, and a decrease in demand for a particular product. Figure 13.5 is an example of what a forecast might look like.

### Simulations and Mathematical Modeling

Making changes in a business system involves time, money, and effort, and can lead to serious problem situations, disruptions in production, and introduction of errors if the wrong changes have been made. Simulation is a technique used to examine the effect that a change in one or more elements in a system will have on the rest of the system—without actually making the changes. It allows the analyst to experiment with various modifications or improvements to learn how each would affect a system without disrupting its normal activities.

The process of simulating a system is done by building a model—a verbal, mathematical description, not a physical construction. It is built by expressing each element in the system as a quantitative value. The simulation is performed by manipulating the values to represent the changes that would occur from modifications introduced into a system. The model is designed

**Figure 13.5.** Forecasting

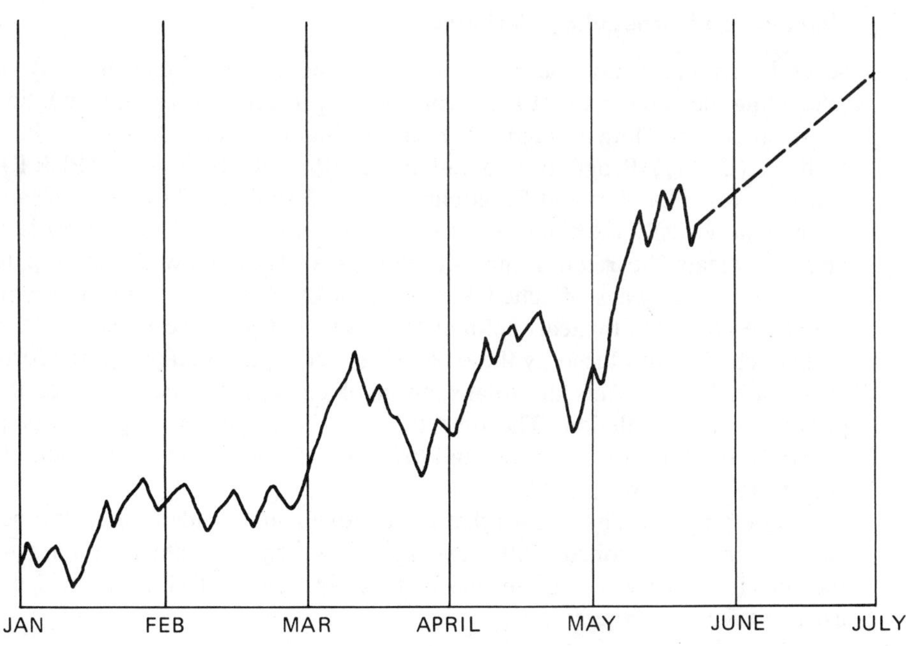

so that a change in one value will produce a proportionate change in the appropriate related values. Performing simulations on the computer increases the practicality and value of the technique and allows experimentation with models and modifications of greater complexity and intricacy.

A firm might use a simulation to experiment with the possibility of installing new equipment—with greater capacity, but at an increase in cost—to see how it will affect such areas as costs, output, number of operators. Simulations can help to predict the effects of raises in material costs, or salary increases. They would also be used to study what would happen if additional work stations were added in a system—before the actual changes were made.

For example, a firm may simulate restructuring its telephone switchboard to find the arrangement that produces the best results. Without simulation techniques, it would be necessary to install a group of telephone lines, establish operator stations, rearrange office facilities, and then check the results.

Using simulation techniques, the problem may be converted into mathematical terms and run on the computer with a simulation program. The effects of the modifications on the other variables in the system (volume of orders processed per hour, operating costs, orders lost because of delays) would be measured and analyzed.

With simulation techniques it becomes feasible to study various modifications and arrangements until the optimum number of lines and work stations are determined. Only after this information has been obtained will plans for implementing the actual changes be initiated.

### Planning and Scheduling Techniques

Several techniques have been developed to find the most efficient way of scheduling the activities that are involved in a project such as implementing a system or building a factory. These techniques—called Critical Path Method (CPM), Program Evaluation and Review Technique (PERT), and Network Planning and Scheduling—are all similar. They are a means of showing visually the time required to complete each task in the project, and to indicate the relationship between tasks. They allow the analyst to compare various ways of scheduling these tasks to find the one that completes the project in the least amount of time and at the lowest cost.

The schedule produced by these planning techniques is shown in the form of a chart. It represents the total time involved from the beginning of the project to its completion. The time that will be spent on each activity is shown in the form of an arrow, and the point of completion of the activity is indicated by a circle.

Very often, several activities that are relatively independent of each other can be carried on concurrently, and are shown by separate time lines on the schedule. A time delay in one of these independent elements will not usually affect the others.

Other activities are interdependent: activity A must be finished before activity B can be started; activity B must be completed before activity C can begin; and so on. A time delay encountered in carrying out one of these dependent activities will retard implementation of all the other dependent activities in that path.

A schedule may contain several paths with dependent activities. The path that represents the greatest time span is called the critical path. It represents the shortest possible time in which the project could be completed. A delay at any point in the critical path will adversely affect the entire schedule, and delay completion of the project. For this reason, the activities in the critical path will be carefully planned and monitored during implementation to minimize disruptions or problems.

Figure 13.6 is a time schedule for producing a technical manual. The critical path for this project is shown by the darker lines.

## EXERCISES

1. What effect did World War II have on the new art of systems analysis?
2. Describe the advantages and limitations of operations research.
3. How do stochastic and deterministic systems differ?
4. List four major sampling methods.
5. Describe a situation in which allocation of resources is an important consideration.
6. What is linear programming and in what kinds of situations is it used?
7. What is the queuing theory and describe a business system situation in which it is used.
8. Describe a major quality control technique.
9. What is forecasting and how it is used?

**Figure 13.6.** Schedule for preparing technical manual

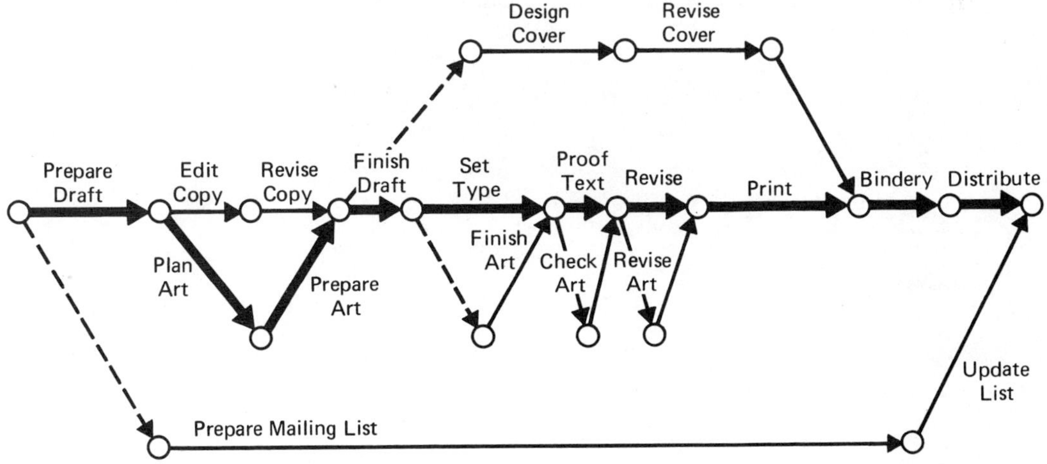

10. Describe the critical path method and give an example of where it should be applied.
11. Imagine yourself enrolling in another college. Prepare a network schedule showing the critical times you will face in making the transition.
12. Visit a business firm and discuss an impending change in its operations. Prepare a schedule showing the critical time relationships involved in the change.
13. Sample the ages of a group of students in your class and determine the average age. Extend the sample to include students in several classes or an entire department.
14. Observe a line or queue at a bank or ticket window. Record your observations on such things as the average waiting time, number who leave without being served, or number of open windows.

# Chapter 14

# Applied Systems Analysis Case Problems

This chapter presents several case histories that illustrate many of the concepts described in this text and expand on the range of business system applications. The varied case histories include local and remote word processing systems, management information systems, order processing, reservation, simulation, and other applications.

Each system is presented in sufficient detail to explain its major aspects, activities, and purposes. Excessive details and information are avoided. A brief sketch describes the business firm and places the system in the context and environment in which it exists and functions. Pictures and flowcharts illustrate important aspects and graphically trace the flow of data throughout the system. Each application includes a system evaluation that discusses the limitations, as well as the benefits, advantages, payoffs, and features of the system.

These case problems have been drawn from actual business situations. They illustrate practical, real life systems, many of which are used by some of the nation's largest and most modern business organizations.

## Risky Insurance Agency

FUNCTION. This unit describes a local word processing system designed to handle text preparation for an insurance agency.

Data input for the system is originated by the insurance agents in the form of spoken dictation or handwritten text. The Word Processing Center is the department in the firm that transcribes this text into finished correspondence. Output from the system consists of such things as personalized letters, memos, quotations, and form letters.

DESCRIPTION OF FIRM. Risky is an insurance agency selling a full line of life, health, and accident policies. It represents about 30 different insurance carriers, and has 25 agents who work out of a local office, selling policies via the mail, the telephone, and personal contacts.

In the course of business, Risky processes a high volume of correspondence, including proposals, price quotations, claims, letters, and memos. Each agent prepares between 50 and 200 pieces of correspondence per week. Dozens of sales brochures and price quotations are mailed out to prospective clients daily. Documents must be neat and accurate and reflect the professional character of the firm. And, of course, the speed with which they are prepared is often essential when processing claims or changes in client's status.

DESCRIPTION OF SYSTEM. Risky's system is organized with the Word Processing Center at its hub. All typing of drafts, finished correspondence, and other text preparation is performed there by employees of the Center. The contents of the material to be typed comes into the Center from the insurance agents. The agents have voice dictation machines available at their desks to facilitate recording the text to be typed. These devices record spoken dictation directly onto magnetic belts. The belts are given to the operators in the Center for transcription and conversion into correspondence, ready for mailing.

The Center is equipped with a magnetic tape word processing machine that is used to transcribe the dictation and produce the finished correspondence. (See Figure 14.1.) This machine can record the keystrokes entered by the operator onto magnetic tape, as well as print out a hard copy of the text. Only the portions of the text needing corrections and revisions need to be re-keyboarded to produce an updated, error-free finished copy.

Figure 14.2 illustrates the general flow of data throughout the system. Agents dictate letters and memos from their desks, including information such as format instructions, name and address of recipient, the text of the memo, and any other relevant details.

Some agents prefer to write out the letter or memo in longhand, and have the operator at the Word Processing Center prepare the finished correspondence from their written dictation.

**Figure 14.1.** Magnetic tape selectric typewriter

Courtesy of IBM.

Several times throughout the day the operators collect handwritten and recorded dictation requests and place them in a box ready for transcription. The magnetic tape word processing machine operator places the belt on a transcription machine which plays back the spoken message. The operator keys the text, and the machine prints out a hard copy draft and records the keystrokes on magnetic tape at the same time. Handwritten text is read visually as it is keyboarded. After the text has been keyboarded, a checking draft or an error-free finished document may be prepared.

Some of the letters and memos used by Risky's agents are standardized, and only the recipient's name and address need to be inserted. In these cases, the agent specifies by its code number which standard form letter is desired. The operator places the magnetic tape containing the text for that letter on the word processing machine and types in the address information. The machine will automatically print out the balance of the letter.

System evaluation. This system has several advantages for Risky. First, centralization of transcription facilities into the Word Processing Center permits specialization of tasks. This means that fewer typists are needed to produce the same quantity of output.

The ability of the magnetic tape word processing machine to store keystrokes greatly reduces the need for retyping, produces error-free drafts, and improves neatness of the finished documents. It also permits the development of a large number of standardized letters.

Another advantage is that the turnaround time for preparation of correspondence has been reduced from several days to a few hours. The ability of the system to print out one or more drafts of a document, without retyping, is an important feature. It allows revisions, corrections, and addi-

**Figure 14.2.** System flowchart—Risky Insurance Company

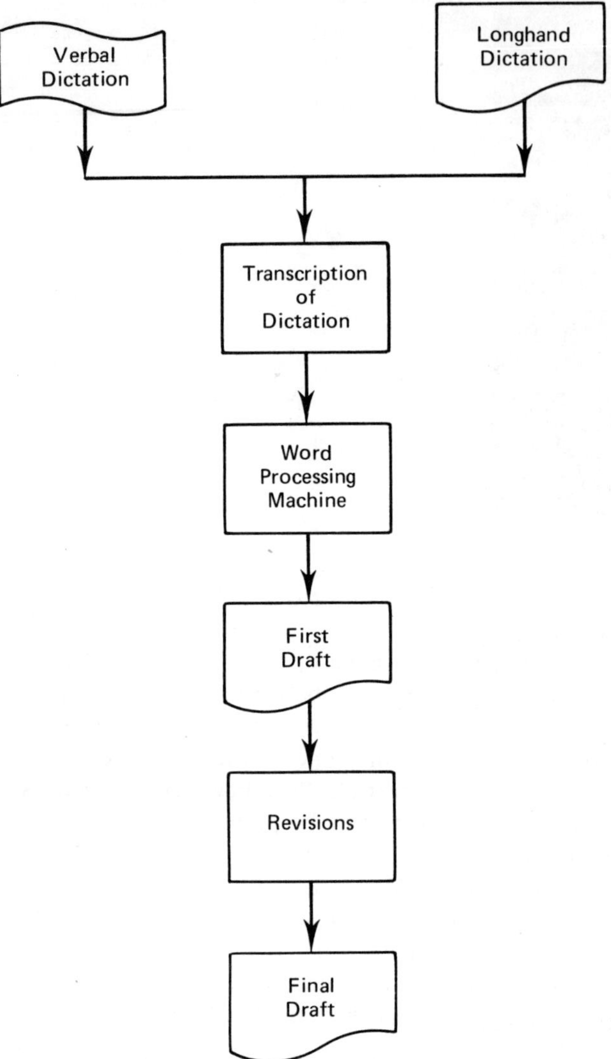


tions to be made easily in a document. It facilitates producing several copies of the same document, either individualized or all the same.

## REMOTE WORD PROCESSING SYSTEM

### Boondocks Farm Equipment Company

FUNCTION. The remote word processing system is designed to allow Boondocks to give responsive, prompt service to its customers, while depending on a traveling sales staff. The system enables salespeople at remote locations to direct the home office to prepare letters, memos, correspondence, and reports.

Most data is input to the system from the traveling sales staff via the mails or the telephone. Other input data comes from the salespeople working in, or visiting, the home office. All text preparation is performed in the Word Processing Center located in the home office. Finished correspondence, letters, and memos are mailed to customers by the Center. Orders received at the Center from the traveling sales staff are forwarded to the sales department for processing.

This system is an example of a remote word processing system with data entry coming from several remote points to a centrally located facility for processing and output.

DESCRIPTION OF FIRM. Boondocks is a leading farm equipment dealer marketing items such as farm implements, fertilizers, seed, and chemicals. A staff of 50 salespeople serves a large geographic area. The sales personnel spend most of their time in the territory to which they are assigned, calling upon accounts. Periodically, they visit the home office for sales meetings.

As the salespeople make calls, they are often asked to mail price quotations, specification sheets, or to answer questions or requests. Since they are in the field for several weeks at a time, they must depend upon the Word Processing Center in the home office to handle these requests.

The sales staff also relies on the Center to route orders, and in some instances to follow up on complaints or delivery schedules. Salespeople also use their remote facilities to forward vital trade information picked up in the field to the home office sales manager for use in planning future sales strategies.

DESCRIPTION OF SYSTEM. Boondock's home office is interested in providing the best possible service to their customers, and efficient, reliable support to their sales staff in the field. A remote word processing and message handling system meets these needs. The system flowchart in Figure 14.3 illustrates the activities it performs.

The heart of the system is the Word Processing Center located in the home office. The Center is designed specifically for preparing letters, memos, and reports, and processing orders and complaints. It is equipped with both remote and local dictation equipment, and a magnetic tape word processing machine.

The word processing machine prints out hard copy documents and stores the keystrokes on magnetic tape at the same time. This greatly facilitates the preparation of reports. The tape can be played out to prepare one or more copies of finished documents with little or no retyping needed for changes, corrections, or additions.

The Center also has a supply of catalogues, bulletins, and folders on Boondock's products. This allows them to handle orders, complaints, and requests for prices promptly and accurately.

**Figure 14.3.** System flowchart—Boondocks Farm Equipment Company

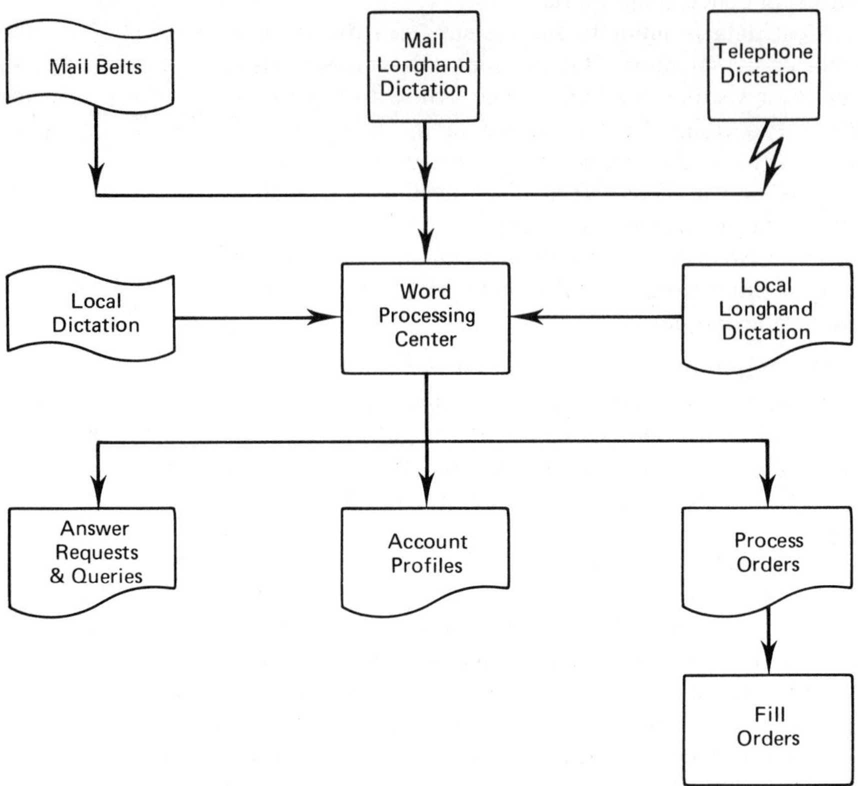


Data is input to the system in four ways:

1. Salespeople in the field telephone the Word Processing Center and are connected to remote dictation machines. They dictate messages or letters over the telephone, which are automatically recorded on magnetic belts.

Periodically, operators in the Center remove the belts and place them on transcribing machines. The contents are keyboarded and stored on the magnetic tape word processing machine. The letters, reports, orders, or complaints are then processed appropriately.

2. Salespeople carry portable dictation equipment in their brief cases or automobiles. (These units are illustrated in Figure 14.4.) This allows them to record messages, orders, or requests on magnetic belts directly in the field. At the end of each day the salespeople mail the belts to the Center for processing. There the contents are transcribed and the documents processed accordingly.

3. Some salespeople mail, or bring personally to the Center, letters, memos, or reports written in longhand. In these instances, the Center tran-

**Figure 14.4.** 274 portable dictation unit

Courtesy of IBM.

scribes the text on the magnetic tape word processing machine and prepares the finished documents.

4. Occasionally, the sales staff comes to the home office for meetings or to catch up on paperwork. In these instances, magnetic belt dictation equipment available in the home office is used. Messages and memos are recorded on magnetic belts and processed by the Center just as though they had been mailed in from the field.

SYSTEM EVALUATION. This system has several advantages for Boondocks. It enables them to centralize their stenographic and word processing services, and thus realize the benefits of a more efficient operation. They have better control over documents. Letters and memos are prepared in a consistent form and style. Orders are processed promptly, and complaints and requests for literature expedited quickly.

Another advantage is that the salespeople have the full resources of the Center at their disposal, even when they are in the field. The system also has benefits for management. Information regarding orders, customers, or new accounts is quickly filtered back to the home office for analysis and referral. This input allows management to make more efficient and responsive decisions with respect to marketing strategy.

## ONLINE ORDER PROCESSING, PRODUCTION CONTROL SYSTEM

### Nuts and Bolts Manufacturing Company

FUNCTION. This system has a triple purpose—it handles the order processing, production control and inventory, and billing and invoicing activities for a medium sized manufacturing firm. The system monitors the the status of orders from the purchase of the raw materials, through production and assembly, and finally to the billing process. It is exemplary of those systems used by manufacturing, fabricating, and assembly companies.

DESCRIPTION OF FIRM. Nuts and Bolts specializes in the manufacture and assembly of small appliances and household hard goods. Most goods are manufactured to order and carry private brand labels or nameplates.

In a typical production cycle, a customer orders goods from Nuts and Bolts to be assembled with its brand name. Nuts and Bolts opens a job number for the order and begins processing. First, the necessary raw materials are ordered and placed in inventory as they arrive. Some of the raw materials are pulled directly from Nuts and Bolts' storage warehouse. When all the necessary raw materials are available, the job is ready for the assembly line. Fabrication usually requires a period of several weeks. Finally, the finished goods are shipped and billed.

Nuts and Bolts requires a system that facilitates efficient and accurate scheduling, order processing, and inventory control. Not only must the proper quantity and kind of raw materials be obtained for each order, but they must be available at the right time to meet assembly schedules. To do this, Nuts and Bolts must know what raw materials are in stock and which are on order at all times. They must know which goods have been assembled, which are ready for shipping, and which have been billed.

DESCRIPTION OF SYSTEM. The system is designed around a computer and several magnetic disk files. Several keyboard terminals located in key departments around the plant are connected online to the computer. (They are illustrated in Figure 14.5.) The terminals in the order department, the production supervisor's office, and shipping and receiving departments allow operators to enter information regarding the movement and status of material and orders. The computer incorporates this new data into the information already recorded on the file. The records on the magnetic file are organized by the job number which was assigned when the order first came in.

Figure 14.6 illustrates the flow of data throughout the system. The cycle starts when an order for a job is received. The order department writes it up and assigns a job number. A computer file for this account is opened on the master disk file under the job number. All data relating to this job will be recorded on the disk file under this number.

A production schedule for the job is developed with the aid of programs

**Figure 14.5.** Online keyboard terminal

Courtesy of Burroughs Corp.

stored on the computer. Another schedule is drawn up to assure that the raw materials needed to produce the finished goods will be available at the right time. Orders for the raw materials that must be purchased are sent out.

Data indicating which materials are removed from Nuts and Bolts own inventory of stock is noted on a terminal in the storage department. A computer program updates the inventory file to maintain an accurate, current picture of the quantity still in stock. As the raw materials purchased outside arrive or are picked from stock, this information is entered from the terminals in the receiving department and incorporated into the file for that job.

At the appropriate points in the schedule, assembly of the goods begins. At various stages, data indicating the current status is entered from the line to update the account file. After the finished goods have been assembled, their shipping dates are recorded from a terminal in the shipping department. Finally, the accounting department uses a computer program to prepare the billing documents from the data recorded under the job number. At any time while the order is being processed in the manufacturing plant, its status can be determined by querying the job number of the account.

SYSTEM EVALUATION. This online system is particularly advantageous for Nuts and Bolts' type of operation. The scheduling programs assure that raw materials will be available at the point they are needed. This facilitates maintaining a more efficient assembly line and avoids delays or interruptions.

Accurate inventory control methods reduce unexpected problems due to items being in short supply. Machines and work stations can be operated

**Figure 14.6.** System flowchart—Nuts and Bolts Manufacturing Company

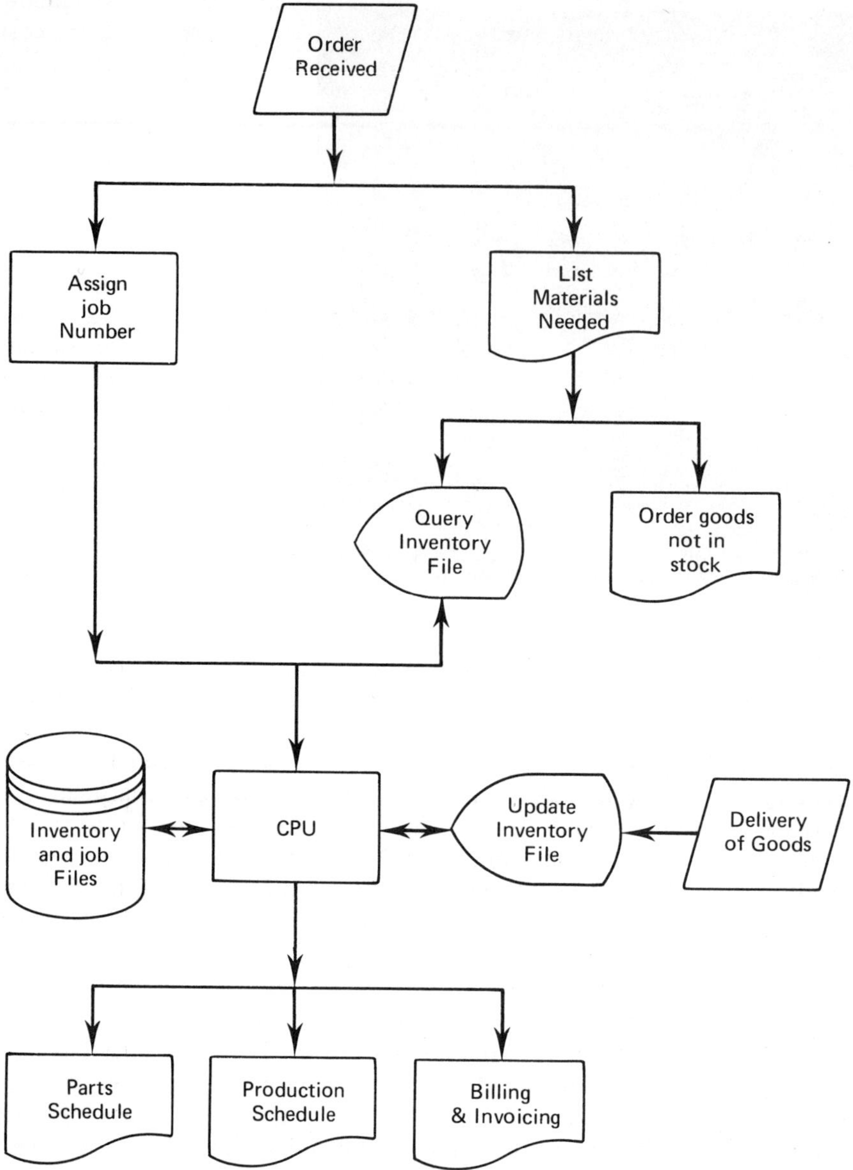


at their optimum level of production efficiency, reducing operating costs. This system results in faster delivery of goods and increases the level of output. Computer facilities expedite the billing and invoicing procedures of finished goods.

The ability to learn quickly the status of any item in the inventory or any job under production enables the firm constantly to monitor schedules and performances and to respond promptly to changes in the environment.

Finally, the system produces a group of useful management reports concerning overall plant operations. These reports give excellent historical data and cost accounting information. The careful charge-back system allows Nuts and Bolts' cost accountants to determine overhead and operating costs with great precision. Ultimately, these advantages lead to improved management decision-making practices.

## OFFLINE RETAILING SYSTEM (SERVICE BUREAU)

### Shirt and Skirt Company

FUNCTION. This system performs the inventory, accounting, ledger, sales, credits, and collections activities for a small garment retailer. Because of its limited size, the Shirt and Skirt Company cannot justify the purchase or lease of its own computer. Instead, it buys these services from an outside company called a service bureau.

Service bureaus are in the business of selling data processing services. They receive the input data from retailers such as Shirt and Skirt via messenger. They process it through their own computer equipment and return the output, such as forms and reports, by messenger. The service bureau is capable of preparing such output as daily balance forms, sales and tax reports, accounts receivables and age analyses, profit and loss statements, and balance sheets. The customers select and pay for the specific types of output they need. Some of the programs used by the service bureau to perform this processing have been written for specific customers for a fee. Others have been written at the service bureau's expense and can be used by any customer, who pays only the fee for the computing time.

DESCRIPTION OF FIRM. Shirt and Skirt Company is a small clothing retailer specializing in men's and women's high-style garments. The company has about 20 employees in full- and part-time positions. The greater percentage of the sales consists of cash transactions, and the remainder are open (charge) account sales. Appropriate records on both types of sales must be kept for tax, audit, inventory, billing, and other purposes.

Most of Shirt and Skirt's success is due to the selection of high-style garments they stock. The buyers rely on up-to-date information regarding sales and the current inventory status in order to identify the lines, colors, and styles enjoying the greatest sales appeal. They select new stock based on this data.

Shirt and Skirt also carries its own accounts, and each month statements must be sent to charge customers. Past due accounts receive letters or notices from the credit manager requesting payment.

DESCRIPTION OF SYSTEM. The system used by Shirt and Skirt has two key elements. One is a special cash register that generates a tape record of all purchases, classified by type of goods, manufacturer, or other code. (This machine is shown in Figure 14.7.) The second element is a service

**Figure 14.7.** Control register

Courtesy of NCR.

bureau. The service bureau processes the data on the tape and generates several reports that are returned by messenger. (Figure 14.8 is a system flowchart of this system.)

Each article of clothing sold by Shirt and Skirt is given a classification code identifying the manufacturer, type of garment, and price. Almost 1,000 different codes may be assigned.

When a cash sale is made, the clerk writes the name and address and other descriptive information in longhand on a cash sales slip. Charge sales are written up on slips of a different color. The slips are placed in the register, which is equipped with several sets of keys. One set is used to enter the classification code, another the amount of purchase, and the third set enters specialized information such as charge or cash sales and taxes. The register also prints out this information on the receipt, which is given to the customer.

As sales are rung up, the register generates a tape record of the data entered for each transaction. This tape captures the data which is used to generate the daily sales journal. This is a complete record of all transactions—cash, open account, taxes, merchandise code—for the day.

At the end of the working day the tapes are removed from the register and sent to the service bureau via messenger. A device at the service bureau reads the data from the tape and converts it into a form that can be processed by the computer. This serves as the principal means of data entry for the system.

A computer program available at the service bureau manipulates this

**Figure 14.8.** System flowchart—Shirt and Skirt Company.

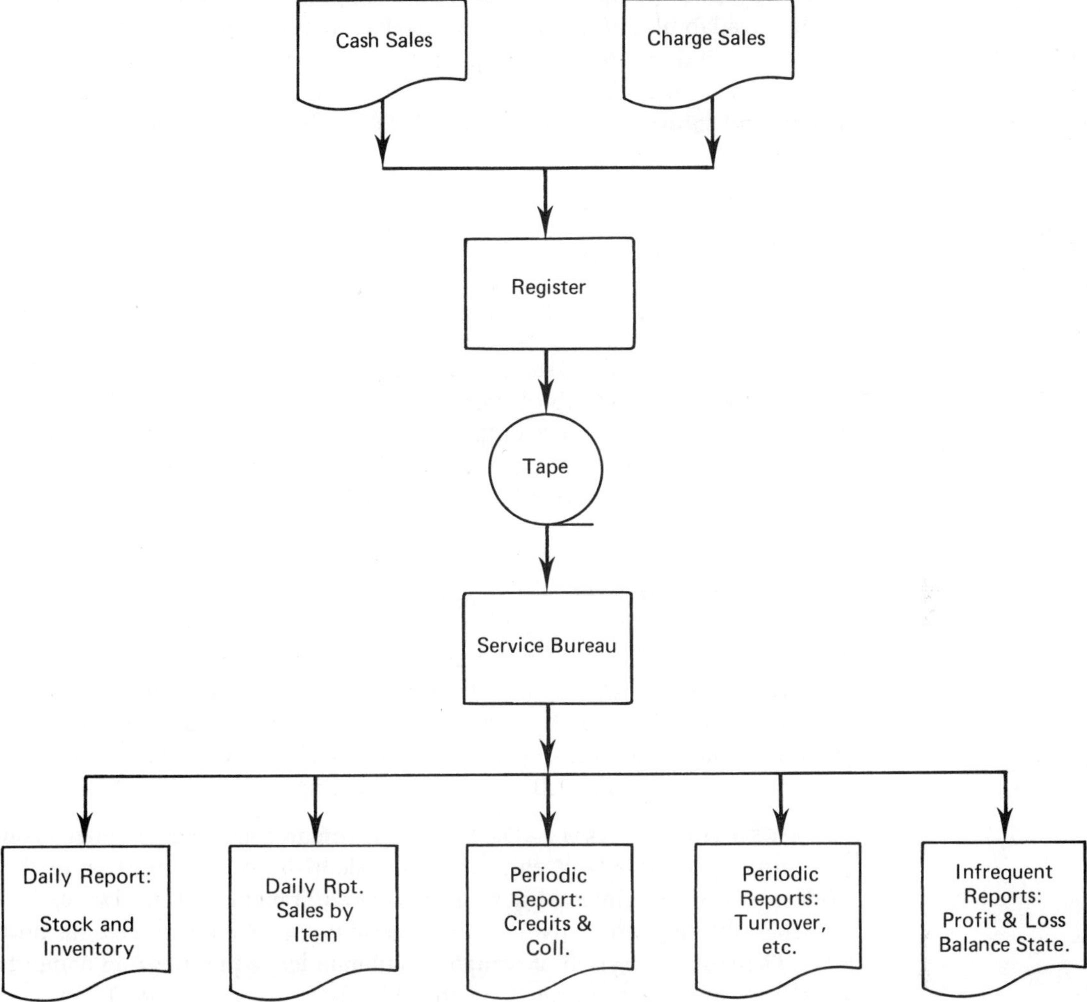


data and prepares a group of reports. The system is programmed to print out daily inventory reports and sales by classification code. Periodically, the program generates turnover, accounts receivables, and age analysis reports.

The credit and collections, and the age analysis reports, are used by the credit manager to facilitate collection of accounts. The inventory reports are studied carefully by the merchandise buyers.

At less frequent intervals, the service bureau prepares profit and loss statements and balance sheets for use in accounting, tax, audit, and management purposes.

SYSTEM EVALUATION. Using a service bureau puts the entire resources of a modern, high-speed computer at the disposal of Shirt and Skirt without the cost of lease or purchase. The classification code facilitates analyzing how each type of garment is selling, and enables the buyers to identify current style trends. The comprehensive inventory and sales reports are used by the buyers to maintain adequate stock supplies of popular merchandise.

The availability of accurate, up-to-date credit reports aids the credit and collection department in keeping the loss from bad debts at a low level. The variety of financial reports generated by the system provides invaluable background information for the decisions that the owner and management must make.

## TIMESHARE AND REMOTE JOB ENTRY SYSTEM

### Da Vinci Research and Development Company

FUNCTION. In this system, online remote facilities provide data processing services for a group of engineers and research scientists. Users enter data, run jobs, and receive output and printouts via terminals located in their offices. These terminals are connected to a computer located at a time share company by telephone lines. No computer is physically located on the Da Vinci premises.

Time share firms sell computing facilities to a company. Their services differ from the service bureau approach (previous case) in several ways. Service bureaus rely on messengers to transport data and output between their site and the client's. Time share companies, on the other hand, rely on a network of terminals located at the user's site and telephone lines as the communications link to the computer. The service bureau provides batch processing only for its customers. The time share firm provides online and interactive facilities, as well as batch processing.

DESCRIPTION OF FIRM. Da Vinci is a research and development company employing dozens of engineers and scientists. A large portion of their work involves experimental designs of new equipment and hardware.

Generally, this work involves two major types of data processing problems. The first is a myriad of complex mathematical equations and computations that must be performed by the scientists and engineers. Results are needed immediately. The researchers often take this type of work home with them to resolve complex problems before the next day's testing begins. The facilities for processing these operations must be portable.

The slide rule and hand held IC calculator suffice for much of this work. But these devices are inadequate for the more elaborate, sophisticated, or repetitive problems that the researchers frequently face. They need the facilities of a high-speed computer for this type of processing. Some problems can be solved with the wide range of mathematical functions available in an interactive mode on the computer. Others are best handled by programs written in the FORTRAN language.

The second category of problems performed by Da Vinci relates to the large amount of data obtained during the development of a new machine or device. This may involve hundreds of observations of equipment as it undergoes extensive testing. This type of data is either punched into cards or recorded on magnetic tape. The results are in the form of lengthy data printouts, or graphs or plots, for proper study and evaluation, or stored on punched cards or magnetic tape.

DESCRIPTION OF SYSTEM. Figure 14.9 illustrates the time share and remote job entry system used by Da Vinci. Two types of terminals are located in the Da Vinci offices. A number of keyboard type terminals, located near the desks of the researchers, provide online facilities for interactive programming. These devices are coupled by telephone lines to the computer at the time share house. From their desks, the researchers can perform a variety of simple and complex mathematical operations. These terminals also provide the ability to execute problems in FORTRAN, as well as other computer languages. (See Figure 14.10.)

Since the system is real time, and online, the data is processed at once and results are returned to the terminal immediately. The terminals are portable and can be taken home by the researchers and connected to the computer via the home telephone. Services are available from the time share firm 24 hours a day, increasing the flexibility of the system.

Several remote jobs entry (RJE) terminals located at keypoints throughout the plant provide batch programming facilities and a variety of output media. These terminals are equipped with a card reader, plotter, magnetic tape reader, and a high-speed line printer. From this terminal data can be entered for processing by the computer on punched cards, magnetic tape, or from the keyboard. Results received back from the computer can be output on the high-speed line printer, the plotter, or recorded on magnetic tape or punched cards for storage. The plotter is capable of producing graphs, charts, or drawings.

The terminals are leased or purchased outright by Da Vinci. At the end of each month, the time share company sends Da Vinci a bill for the amount of computer processing time used. Costs for lease or purchase of computer equipment is avoided.

**Figure 14.9.** System flowchart—Da Vinci Research and Development Company

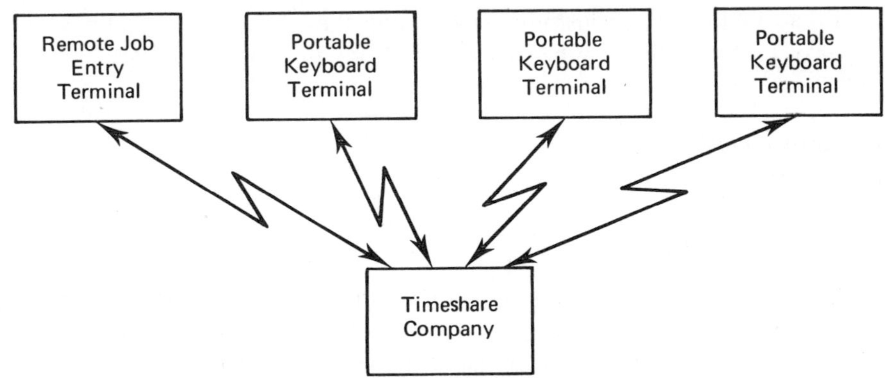

**Figure 14.10.** Timeshare system

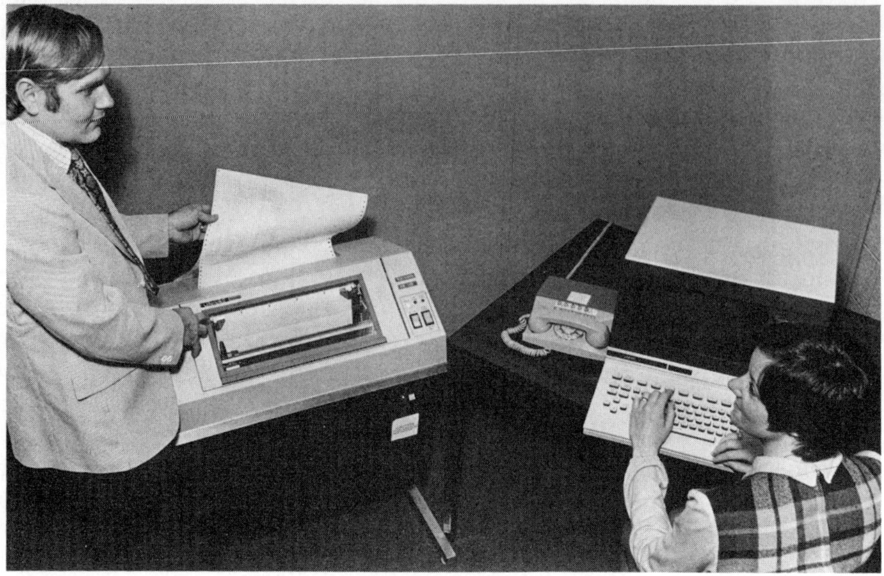

Courtesy of Sperry Univac, a division of Sperry Rand Corporation.

SYSTEM EVALUATION. The system provides Da Vinci with the full range and power of a high-speed computer, without the expense of operating, leasing, or purchasing its own equipment. The two types of terminals in use conveniently handle the range of problems facing the researchers at Da Vinci, and provide a variety of output media.

This system has a considerable degree of flexibility. Terminals are easily relocated in the plant and can be conveniently taken home. Management can modify the arrangement easily, adding new terminals as needed or installing additional output devices.

There is no problem with overloading the system, since computer time can be purchased from several time share firms at once. Da Vinci is not involved with the problems of equipment limitations, such as obsolescence and maintenance. Time share firms make a variety of languages and programs available to the customer, increasing the range of applications and types of problems that can be solved right from a researcher's desk.

## ONLINE, REAL TIME INVENTORY, AND POINT-OF-SALE SYSTEM

### Slick Paper Company

FUNCTION. This example illustrates an online, real time inventory system that provides a large paper merchant with up-to-date information on its inventory of stock. The system also supports a group of point-of-sale terminals, relaying data regarding sales transactions to the home office as they occur. The system maintains the master file, listing all goods in the inventory, their quantity, price, or supplier. In addition, the system provides management with inventory turnover, order points, and billing and accounting reports.

Systems such as this can be used by large retailers, wholesalers, and jobbers. Their distinguishing feature is the real time capability. This means that as items are sold, or new goods placed in stock, the master inventory file is updated immediately. Accessibility to this current information gives salespeople, buyers, or managers an important advantage when completing transactions, such as taking orders or purchasing stock.

DESCRIPTION OF FIRM. Slick Paper Company is a large paper supplier with branch sales offices and warehouses located in major cities in several states. Slick stocks a great variety of paper styles in many different widths and colors. This creates a sizable and involved warehousing and inventory job. To complicate the situation further, the company often accepts orders for goods in one office but ships them from the stock available in another office.

Slick's sales staff calls upon printers, publishers, advertising agencies, and others. These salespeople must know the exact quantity of goods available in stock, especially when a sizable amount of paper is involved in the order. Insufficient supply of an item in stock could delay the shipment of goods to the customer, resulting in the loss of goodwill and, possibly, future sales.

Many sales are also made at the branch offices by customers who personally pick up goods at the will-call counters. These transactions must also be entered into the inventory file immediately.

DESCRIPTION OF SYSTEM. A large computer system with disk storage capability is located in Slick's main office. The master inventory files are recorded on the disk media. A series of remote online terminals are connected to the computer either by telephone lines or direct wire. These terminals are located at sales desks, order, warehouse, shipping, and receiving departments of all the branches. (Figure 14.11 illustrates the arrangement of the system.)

The real time capabilities of the system mean that transactions are recorded as they occur. As new stock is added, or items are sold, information regarding these changes is immediately entered into the inventory file. The status of this file can be queried from each terminal. This allows salespeople in the field to call the order desk and immediately learn the quantity in stock, price, or other information, of any item. As they place orders, the quantities are deducted from the totals available.

To facilitate transactions at the will-call counters, Slick has installed a group of CRT point-of-sale terminals at each order desk and in the stockroom. (See Figure 14.12.) Data from these transactions are entered into the master file as they occur.

At weekly and monthly intervals, the system is programmed to prepare management and inventory reports. One report lists each item in stock and

**Figure 14.11.** System flowchart—Slick Paper Company

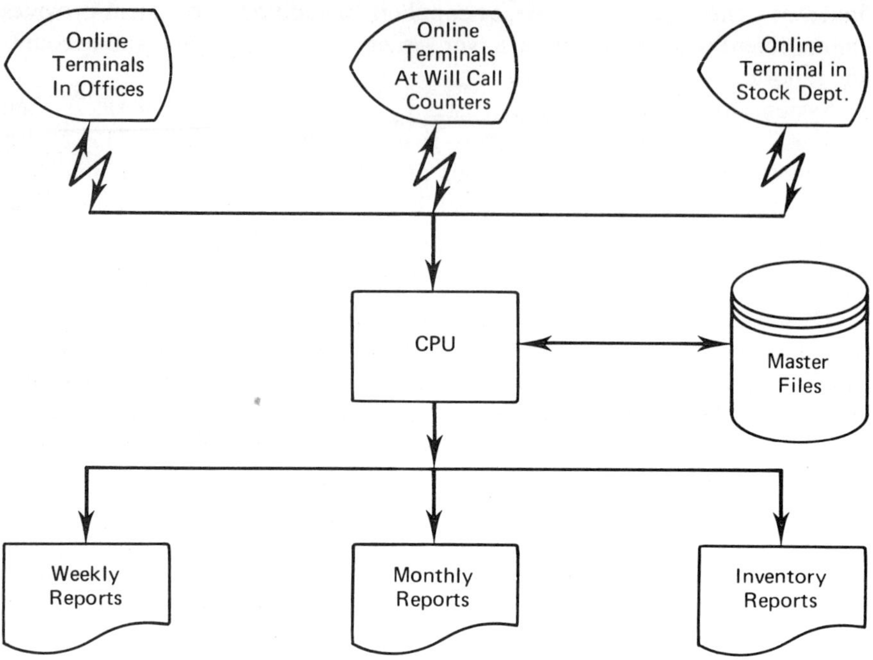


**Figure 14.12.** Point-of-sale terminal

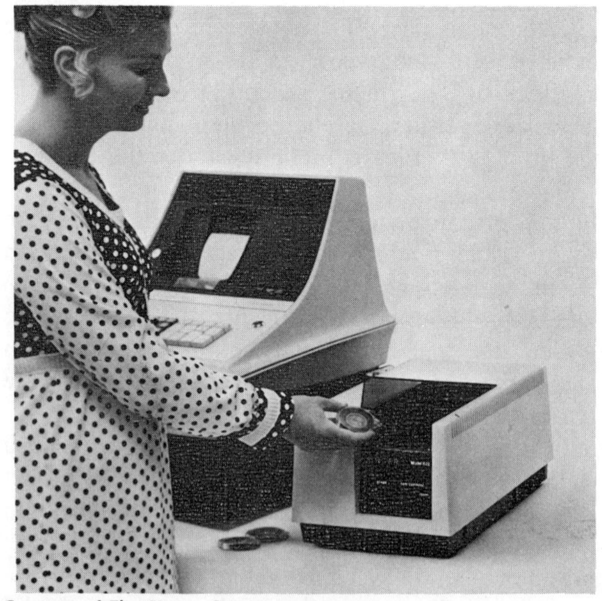

Courtesy of The Singer Company.

indicates its turnover ratio. Another report flags each item in short supply. Additional reports list sales by salesperson, type of goods, branch, department, and vendor.

SYSTEM EVALUATION. The system has been very beneficial for Slick. The instant accessibility to accurate stock information has greatly increased Slick's ability to land large sales orders and has improved its performance in a very competitive marketplace. Errors in maintaining inventory records have been greatly reduced. This has resulted in less instances where goods are unexpectedly out of stock, encouraging customer goodwill and dependency.

Service at the will-call counters is prompt and efficient. The point-of-sale terminals have sped up processing these kinds of sales. The reports prepared by the system encourage better management practices. The reports which flag items in short supply give valuable information to buyers who must determine the popular items and the vendor lines which move the fastest.

An excellent picture of sales performance is generated by the report that shows the activity of each salesperson, branch, and product line. Information on discontinued merchandise and outdated colors and sizes can be easily gathered and disseminated, encouraging their prompt and profitable disposal.

## ONLINE RESERVATION SYSTEM

### Sleep Tight Hotel Chain

FUNCTION. This system is designed to facilitate processing hotel room reservations. It is typical of business systems used by hotels, motels, airlines, and automobile rental organizations to manage reservation activities.

The purpose of this system is to maintain a master file describing room accommodations available at all the facilities in the chain. This file can be accessed from remote terminals located in the reservation offices of the various hotels in the chain. Provisions are also available for outside ticket and travel agencies to make reservations for accommodations.

The concept of online, real time reservation systems is similar, regardless of the type of firm involved. The master data file contains all relevant information describing the services being sold. For hotels and motels this would include a description of the room accommodations—number and type of beds, prices, dates available. Airlines would include flight schedules, seats available on each flight, price.

The system must maintain the master file to reflect the changes in availability as they occur. As reservations are made, these units are removed from the master file. As new units become available, they are added to the list.

In some arrangements, the master file can be accessed, queried, and updated from the remote terminals. Other arrangements only permit data query from the remote stations—all maintenance and updating are performed from the central location.

Each of the 30 facilities in the chain is equipped with its own remote terminal, online to the central computer. (See Figure 14.14.) Requests for line is answered by operators at the firm who access the master file via terminals and relay reservation information between the outside agency and the computer file.

DESCRIPTION OF FIRM. The Sleep Tight Hotel Chain operates 30 hotels and motels located in several major cities in different states. Each facility has different types of accommodations available, including singles, doubles, rooms for families, and suites.

The rates for the accommodations vary depending upon the season, size of room, number of beds, and the location of the hotel or motel. Some of these facilities grant discounts to selected commercial firms who are regular and frequent clients. Names of these accounts are listed on a special file.

Reservations for all accommodations can be made directly from any of the 30 members of the Sleep Tight chain. Reservations can also be made via phone from outside agencies.

DESCRIPTION OF SYSTEM. Figure 14.13 illustrates the reservation system used by the Sleep Tight chain. A large computer is located in the administrative offices at Sleep Tight's headquarters. The master file of accommodations available at all units in the chain is recorded on a magnetic disk storage device connected online to the computer.

**Figure 14.13.** System flowchart—Sleep Tight Hotel Reservation System

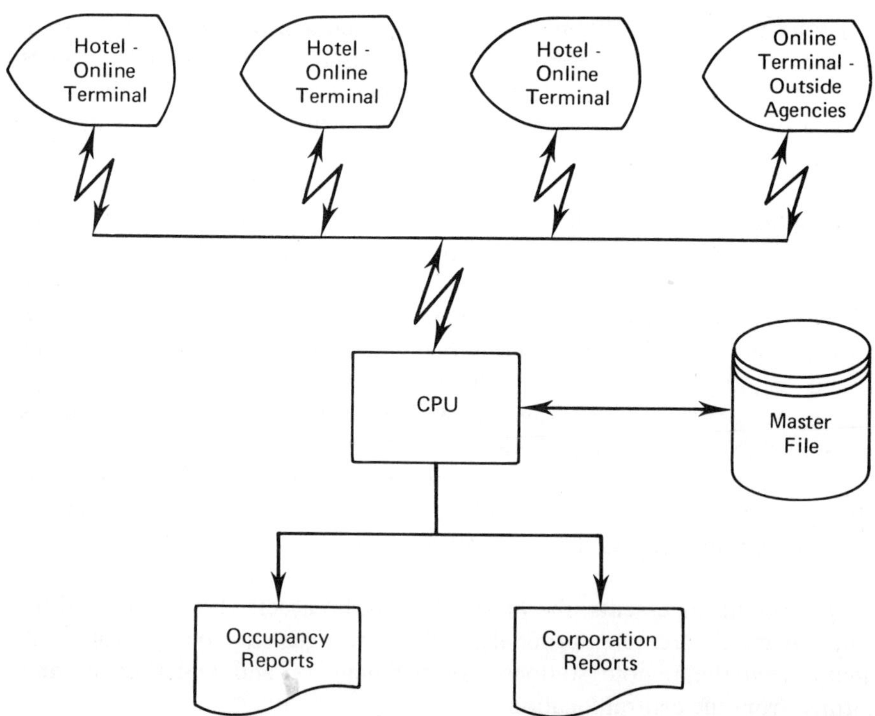

In many instances, inward WATS lines are installed to allow outside ticket agencies and travel agents to place reservations. This no-toll telephone reservations come to the terminal operators from customers either in person, over the phone, or by mail. The operator determines the type of accommodation needed, location of facility, and dates. The computer is directed to access the master file and display the accommodations available for that facility on those dates. The accommodations that most closely meet the request are selected.

If the original request is not available, the computer is directed to display alternate selections—different dates, another combination of beds and rooms.

After the customer has made a selection and purchased the accommodation, the data is keyed into the terminal and a hold is placed on that particular accommodation. A written confirmation is handed or mailed to the customer, showing the date, price, and other information.

Another terminal at the home office is available for processing reservations from outside agencies which come in through the mails or via the WATS lines. The operator keyboards the request into the system and gets an immediate display of available accommodations. Verbal confirmation is made immediately over the telephone and a written confirmation mailed to the agent.

Periodically, the computer is programmed to print out several reports on the activity of the various hotels and motels in the chain. These reports indicate such things as average length of stay, percentage of time each type of accommodation is utilized, number of reservations cancelled, and number lost due to unavailable facilities.

EVALUATION OF SYSTEM. Sleep Tight realizes several advantages from using a real time computerized reservation system, instead of telephone operators or mail facilities alone. The direct costs involved in operating the

**Figure 14.14.** Online terminal

Courtesy of Hewlett-Packard.

reservation system itself are less than those in the non-computerized arrangement.

The efficiency of the system results in maximum occupancy of each facility in the chain. The accommodations available in a given facility may change rapidly, depending upon such factors as weather and travel conditions. Cancelled accommodations are immediately reentered into the master file to be resold. Losses from unused accommodations are less when information regarding current status is immediately available to ticketing agencies.

Since the customer receives a printed confirmation indicating the number of days of stay, type of room, cost, and other details, errors and no-shows are reduced.

## SIMULATION AND MODELING SYSTEM

### Best Banana Shippers

FUNCTION. This system facilitates loading and scheduling a fleet of oceangoing vessels which ship a cargo of bananas and other tropical fruits from South America to the United States and parts of Europe. The system is used to program the best combination of fruits for a given cargo and the optimum departure and arrival schedules. The system includes a modeling program which allows Best Banana to experiment with alternative loading patterns and itineraries. The modeling program is also used to schedule future planting selections and times.

DESCRIPTION OF FIRM. Best Banana maintains a fleet of oceangoing vessels which ship bananas and other tropical fruits from several major ports in South America. The fleet contains 12 ships varying in tonnage and cargo-carrying capacity. The cargoes are shipped to various ports in the United States and Europe, where Best Banana operates several dockside unloading facilities. After unloading, the goods are sent via rail or truck to markets in major inland cities.

The problem is to find the most profitable schedule for shipping the produce to market. This is a complex situation with many other related factors that must be considered. Obviously, the shorter the run, the lower the shipping costs. But often many ships are at sea at one time when crops come to harvest, limiting the flexibility offered by their varying capacities. The ships make several stops, taking on various fruit cargoes and unloading others. In addition, the needs and schedules of the customers buying the fruit must be met. And, of course, at all times the specter of spoilage of the cargo items adds a degree of complexity.

All these factors create a complicated, intricate scheduling problem. In the past, a traffic manager was assigned to this task. But with the myriad of factors to be considered, the unexpected difficulties that arose, bad weather, and human errors, this job became too complicated to be handled efficiently by people. Computerized facilities were indicated.

Another of Best Banana's problems is to determine the best combination of crops to be grown and the optimum times to plant them to meet customer's projected needs at harvest time. These planting decisions must be made many months before the orders come in, and, of course, well before the harvest. Factors such as weather or variations in productivity must be considered. Harvesting times and degree of ripeness of produce must be coordinated. Fruit picked too soon will not ripen properly. Fruit picked too late will spoil in transit. Errors in judging any of these factors will cost Best Banana money.

DESCRIPTION OF SYSTEM. Figure 14.15 illustrates the system developed by Best Banana. A computer located at its headquarters is programmed to generate schedules and simulations. A planning program develops the itinerary schedules and a linear programming technique is used to determine the best combination of fruits for the cargoes of the various ships.

A modeling program of the entire shipping and cargo operation gives Best Banana the capability of simulating many of the effects of varying conditions, and experimenting with loading arrangements, routes, and schedules. The modeling program is also used to develop several planting schedules which consider all possible changes in conditions and factors.

The system is designed to produce several reports. One indicates the status of all orders in the house. Another lists the cargo and destination of each vessel in the fleet. Other reports and tables show departure times and ports of embarkation and arrival. The planting schedule is carefully defined in detail by another report. Status rosters are prepared for each dock to

**Figure 14.15.** System flowchart—Best Banana Shippers

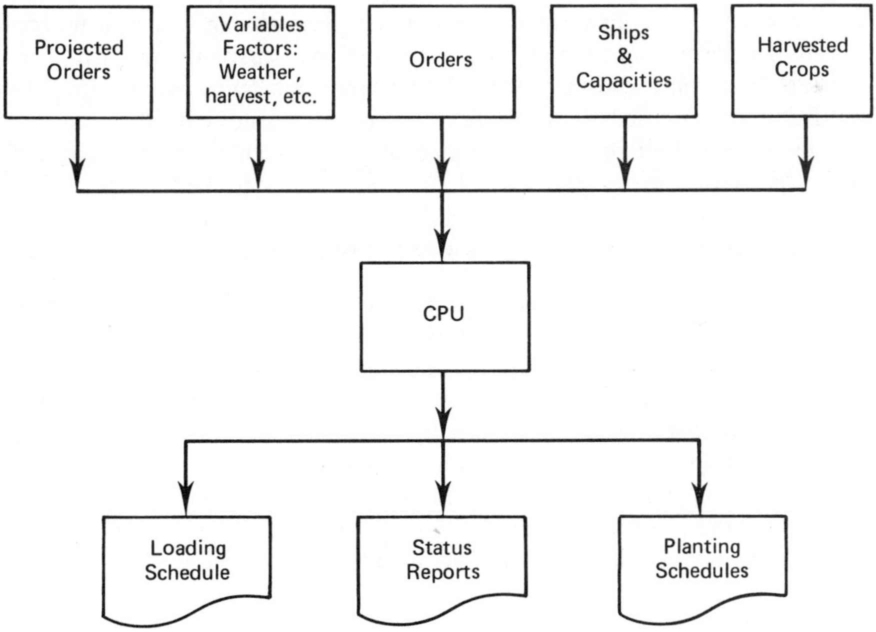

indicate when shipments are due, how the cargo should be processed, and what surface transportation should be utilized to take the fruit to its final destination.

EVALUATION OF SYSTEM. This computerized system gives Best Banana several advantages over a manual scheduling program. Goods are shipped faster and arrive in better condition at the customers' facilities. Goods do not remain on the docks waiting for ships to arrive, risking spoilage due to scheduling errors or poor planning.

Increased efficiency in loading cargoes has led to lower operating costs and larger profit margins. The status rosters for the docks and other work stations facilitate planning work schedules for personnel and equipment.

Finally, the schedules generated for planting crops are extremely valuable. They have increased Best Banana's ability to make sound decisions regarding future harvests and to meet projected markets. They allow Best Banana to study and analyze alternative planting schedules and crop mixes.

## INFORMATION RETRIEVAL SYSTEM

### Big Town *Gazette*

FUNCTION. This system provides the Big Town *Gazette* with an efficient information retrieval and filing system. It is used to maintain files and archives for research by reporters, editors, and other newspaper personnel.

These types of files, called morgues, are maintained by many newspapers. A morgue is a file of newspaper articles, pictures, and clippings on important people, events, and happenings. Some newspapers also maintain a file composed of a permanent copy of each edition of the paper they publish. These files are important research media for writers, reporters, and editors. They are frequently queried to check data, names, and historical information.

This information retrieval system enables researchers to locate specific items quickly from the millions of pictures, clippings, and articles in the morgue. Similar information retrieval systems are used by schools, colleges, and libraries to catalogue, index, and file information.

DESCRIPTION OF FIRM. Big Town *Gazette* is a large metropolitan newspaper, employing dozens of editors, copywriters, reporters, and news analysts, who prepare the copy for the daily editions. One of their duties is to check the accuracy of news items and to research background information on current stories. They frequently review dozens of clippings, articles, or old newspaper pages when pursuing a news story.

Big Town *Gazette* has been published for over 50 years and is proud of its morgue and extensive file of old newspaper editions. In these files can be found references to all the important people, events and viewpoints of the last 50 years.

Description of system. Figure 14.16 illustrates the major aspects of the information retrieval system. There are two essential elements involved: the microform files and the computer based indexing system.

Each day's editions are copied in full on microfiche cards. (See Figure 14.17.) Each card contains a microfilm record of up to 80 pages. Other miscellaneous clippings and pictures are copied on aperture cards (Figure 14.18). These are similar to microfiche, except that each image is on a separate card. Both of these files are stored in filing cabinets in sequential order. The cards can be collated and sorted by conventional unit record machines.

Each day Big Town's librarians copy pages from the current edition onto microfiche and have clippings and pictures copied onto aperture cards. These are merged into the master files, ready for access by the library clerks.

To view either the aperture or microfiche cards, the microform is pulled from the file and placed in an appropriate reader (Figure 14.19). The readers enlarge the images back to their original size for easy reading.

**Figure 14.16.** System flowchart—Big Town Gazette

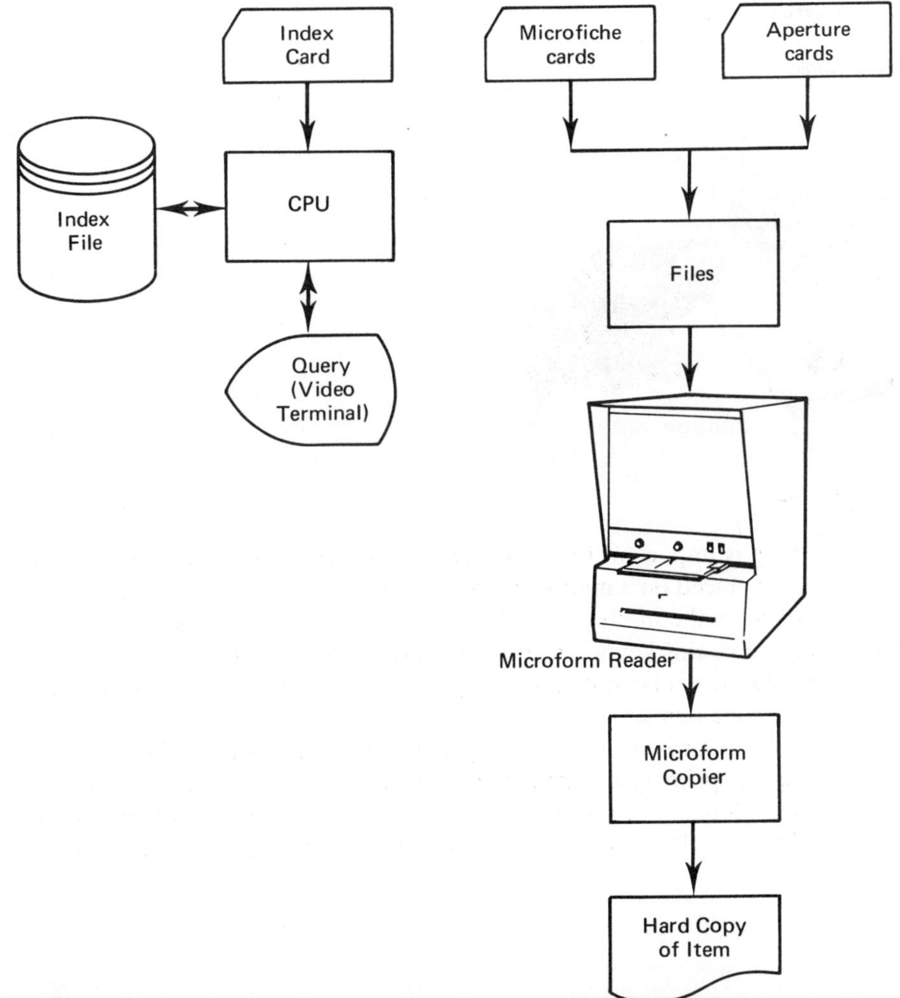

**Figure 14.17.** Microfiche cards

Courtesy of the Eastman Kodak Company.

**Figure 14.18.** Aperture cards

Courtesy of the Eastman Kodak Company.

Indexing is important in this system. An index card is prepared for each item as it is placed on a microform record. This card is fed into the computer for inclusion in the master index file, which is stored on magnetic disk. Each index card contains key words describing the items, called descriptors. Index cards containing related or allied key words are also included in the file to cross-reference items.

To look up an item, a reporter goes to the librarian and fills out a card indicating the primary descriptor and related key words. A clerk enters this information into a video terminal connected to the computer, which accesses the master file. All of the related items in the file are displayed on the screen,

**Figure 14.19.** MISI microform reader

Courtesy of Micro Information Systems.

together with filing reference numbers. The reporter copies these down and requests the appropriate microform records from the librarian.

For example, to obtain information on South American rubber plantations, the reporter would list the key word "rubber," followed by the modifiers "plantation" and "South America." The computer then searches the files and displays the reference numbers of all related documents in the file. The reporter selects the appropriate records.

If a permanent copy of the record is needed, the microform record is obtained from the librarian, then placed in a microform copying machine. This device enlarges the image to life size, and prepares a permanent, hard copy of the original clipping or article. The microform record is then returned to its place in the file.

EVALUATION OF SYSTEM. This system has been very advantageous for the staff members of the Big Town *Gazette*. It enables them to search important records quickly, easily, and thoroughly. The comprehensive index system of related key terms facilitates expanding searches into allied areas.

Since microform records can be copied, the risk of loss and damage to paper originals is avoided. A backup copy of all records is kept in an archival storage file in case a microform or aperture card is lost or destroyed.

The system is an economical method of storing large volumes of visual and graphic data. Considerably less space and filing equipment are required than to store copies of the newspapers and documents in their original form.

## MANAGEMENT INFORMATION SYSTEM (MIS)—DATA BASE

### Tweed City Hall

FUNCTION. The system in this example is an online management information system (MIS) serving the accounting, management, budget, and data processing needs of an entire city. It performs the processing for the court system, the business licensing, health, police, fire, roads, and motor pool departments.

The system generates reports and data used by the city council and city managers in their decision-making capacities. It handles the payroll and fiscal accounting procedures for the city employees. It maintains files for several city departments and provides online access to these files.

DESCRIPTION OF CITY. Tweed City is a medium sized town, about 600,000 population, with its own police, fire, and health departments. The municipal government issues business licenses, collects taxes, maintains an animal shelter, road maintenance department, and a municipal court system.

Each business day a considerable number of transactions occur between the city agencies and departments and the citizens of Tweed City. Licenses are issued, court cases processed, fines collected, stolen property recovered.

All of these activities generate a massive amount of data, which must be processed and recorded in various ways. Some of the data is needed immediately for processing requests, making decisions, answering questions, and so on. Some is used as background data in future planning processes to guide expenditures and allocation of city resources. Other data is used for tax purposes, and some is of an historical nature.

DESCRIPTION OF SYSTEM. The focal point of Tweed City's system is a large computer and magnetic disk storage system which holds the city's data base. (See Figure 14.20.) The data base is a collection of master files composed of the records from the various operating departments in the city. The files are updated from data generated as transactions occur in the various departments. Online video display terminals (CRTs) are located in certain city offices, such as the police, fire, business license, and court facilities. Specified users access the data base from these terminals to retrieve information, and, in some instances, enter new data.

Two types of output are generated by this system. One is the online query service available via the video terminals in the various departments. The second type of output is hard copy printouts of reports and documents.

A large number of reports are generated by the system for the various

**Figure 14.20.** System flowchart—Tweed City Hall

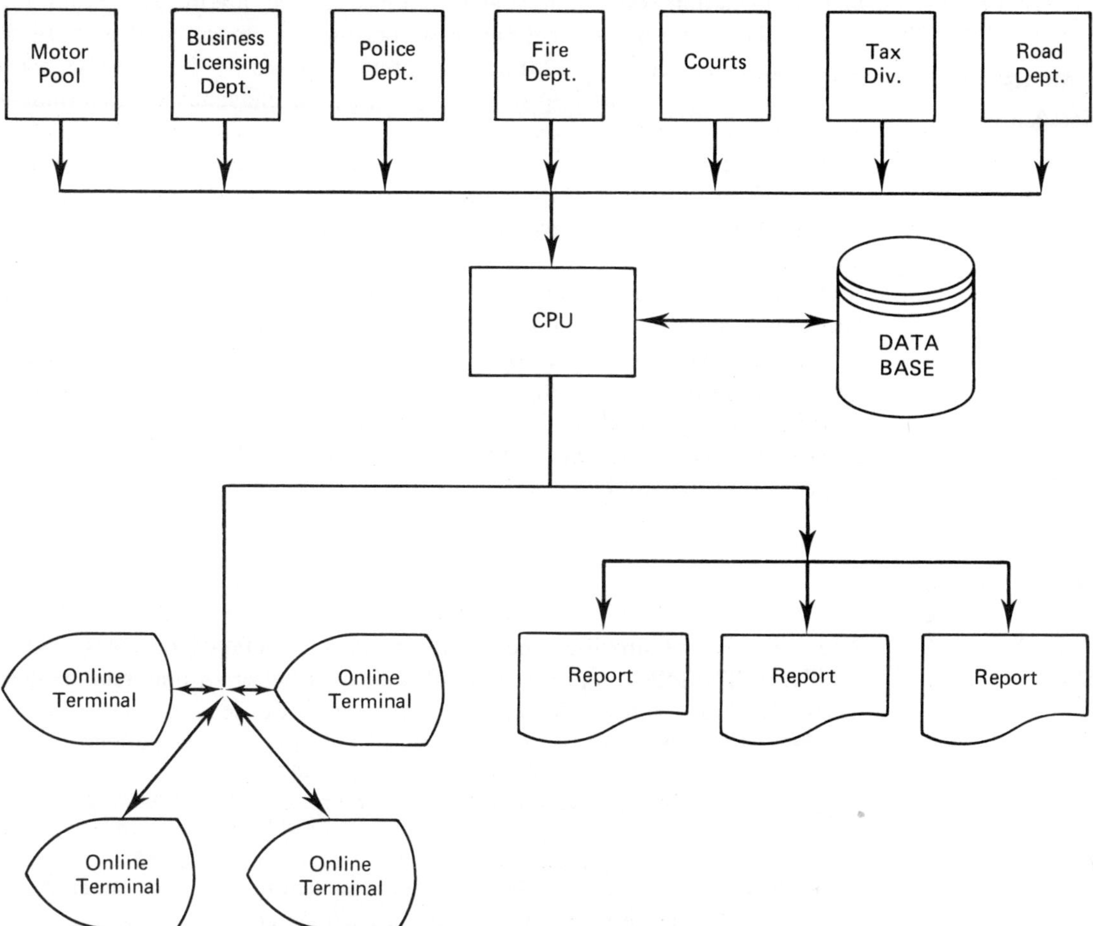


agencies. The value of these reports is often directly in relation to the quality of the data used in their preparation. Reports based on accurate, current, and comprehensive resource data help the agencies respond efficiently and intelligently to the demands placed on the system.

Reports on the court case load are generated for the probation department to facilitate scheduling of rooms and personnel and to anticipate trends and future needs. Studies and comparisons of changes and variations in the tax base are used when calculating the next year's tax rate. A schedule listing the dates for servicing the motor vehicles in the city-owned fleet is prepared for the motor pool. This helps to coordinate maintenance and improve efficiency.

The speed and ease of data retrieval is also an important factor in many instances. This is particularly true in many of the activities of the police and fire departments.

EVALUATION OF SYSTEM. This system illustrates an efficient management information system. Since all major city services are integrated into the system, the government functions as a coordinated, interrelated unit. Each department has access to all the information gathered by the other agencies as well as its own. This improves the quality and efficiency of government and reduces operating costs.

The reports generated by the system represent an accurate and up-to-date picture of the activities of the various departments. This provides a sound, reliable base for the governing agencies to rely on and improves their decision-making practices.

The accessibility of the information in the data base enables the governing agencies to serve the public faster and more efficiently.

On the other hand, of course, there is always the possibility of abuse. The information in the data base must be protected so that it is always used in a proper, legal, and appropriate manner. A system such as this is successful only as long as serving the best interests of the public is always its prime objective and goal.

## EXERCISES

The cases in the previous section illustrate a variety of business system problems. The following exercises will help you better to understand these cases. Do these exercises for each of the cases in the section.

1. The systems described require a variety of pieces of documentation. Prepare a different piece of documentation for each system.
2. Modify the flowchart to change, alter, or improve the system. Document the changes.
3. Suppose the changes proposed above were to be implemented. What steps would you take to see that they were implemented in an orderly manner?
4. Assume the business volume handled in each system were to double. What changes or modifications in procedure, policies, methods, personnel, and equipment would be necessary?
5. Assume the business volume handled in each system were to be reduced by half. What changes or modifications in procedures, policies, methods, personnel, and equipment would be necessary?

# Index